Invertebrates of Central Texas Wetlands

Invertebrates of Central Texas Wetlands

Stephen Welton Taber

and

Scott B. Fleenor

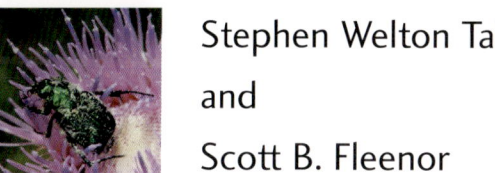

Texas Tech University Press

This book is typeset in Bitstream Arrus. The paper used in this book meets the minimum requirements of ANSI/NISO Z39.48-1992 (R1997). ∞

Library of Congress Cataloguing in Publication Data
Taber, Stephen Welton, 1956–
    Invertebrates of central Texas wetlands / Stephen Welton Taber and Scott B. Fleenor.
        p.   cm.
    Includes bibliographical references and index.
    ISBN 0-89672-542-1 (cloth : alk. paper)—ISBN 0-89672-550-2 (pbk. : alk. paper)
    1. Invertebrates—Texas. 2. Wetlands—Texas. I. Fleenor, Scott B., 1962– II. Title.
    QL365.4.U6T33 2005
    592'.09764—dc22
                                                                2004010712
ISBN-13 978-089672-542-3 (cloth)
ISBN-13 978-089672-550-8 (paper)

04 05 06 07 08 09 10 11 12 / 9 8 7 6 5 4 3 2 1

Printed in China

Texas Tech University Press
Box 41037
Lubbock, Texas 79409-1037 USA
800.832.4042
www.ttup.ttu.edu
ttup@ttu.edu

To Harvey Soefje

Contents

Preface

The wetlands of central Texas, or, more precisely, south-central Texas, lie on both public and private properties, and so we thank Sheriff Glen A. Sachtleben and Mr. Garry Henderson of the Gonzales County Appraisal District for guiding us to private landowners when we began our study. Permission to enter and explore was kindly granted by Mrs. Evelyn Pettus, Mr. Will Soefje, and the late Mr. Harvey Soefje, to whom this book is dedicated for graciously allowing us free rein in Soefje Swamp. We thank Mrs. Corenna Walker and Mr. Dale Walker for introducing us to Mr. Soefje in the field.

Access to the publicly owned wetlands requires no permission, because these lie within the boundaries of Palmetto State Park. Nevertheless, a permit is required to collect plants and animals for purposes of identification, so we thank Dr. David Riskind of the Texas Parks and Wildlife Department for issuing Scientific Study Permit No. 21–01. Park Superintendent Mark Abolafia-Rosenzweig alerted us to the presence of local species that might otherwise have been overlooked, apprised us of relevant developments, including flowering periods, rainfall, temperature, and flood status, and provided answers to many other questions. Park Manager David Allen contributed specimens of a spectacular predatory grasshopper that we encountered for the first time during our studies in the Ottine area.

Academic scientists, private specialists, and the staffs of state, federal, and private museums offered aid and advice in the identification of plants and animals. For this we thank Dr. George E. Ball of the University of Alberta; Dr. Denton Belk of San Antonio, Texas; Dr. Yves Bousquet of Agriculture and Agri-Food Canada; Dr. George Byers of the University of Kansas; Dr. John Capinera of the University of Florida; Dr. Scott P. Carroll of the University of California at Davis; Dr. John A. Chemsak of the University of California at Berkeley; Dr. Ted Cohn of San Diego State University; Dr. Atilano Contreras-Ramos of the Universidad Autonoma del Estado de Hidalgo, Mexico; Dr. Patrick De Clercq of Ghent University (Belgium); Torsten Dikow of the University of Rostock (Germany); Dr. Terry Erwin of the United States National Museum; Mr. Lawrence Forcella of godofinsects.com; Dr. Herb Levi of Harvard University; Dr. Robert E. Lewis of Iowa State University; Dr. Steven W. Lingafelter of the USDA; Dr. Jeffrey A. Lockwood of the University of Wyoming; Dr. Joel W. Martin of the Los Angeles County Natural History Museum; Dr. J. E. McPherson of Southern Illinois University; Dr. Alfred F. Newton of the Chicago Field Museum; Dr. Felipe Noguera of the UNAM Estacion de Biologia Chamela at Jalisco, Mexico; cartographer Cheryl O'Brien of the United States Geological Survey; Dr. John Pinto of the University of California at Riverside; Dr. Michael Pogue of the USDA; Dr. Karen J. Reed of the United States National Museum; Dr. Rowland M. Shelley of the North Carolina Museum of Natural Sciences; Dr. Ned E. Strenth of Angelo State University; and Dr. Thomas J. Walker of the University of Florida. Dr. James Dixon of Texas A&M University kindly gave permission to use a map that formed the basis of Map 1-1, and Cheryl O'Brien and Dr. David Riskind alerted us to the fact that USGS and Texas State Park maps lie in the public domain. Dr. Riskind also read the manuscript in its early stages and provided helpful criticism.

We close these acknowledgments with homage to the work done by two early field biologists. In the early years of the twentieth century the botanist Edwin Robert Bogusch conducted a survey of plants in an area near our study sites, but we did not see this area (Bogusch 1928, 1930). This survey was followed in the middle of the century by Gerald G. Raun's zoological survey of mammals, reptiles, and amphibians (Raun 1958, 1959). Raun regretted that a study of his favored subjects had not been undertaken by Bogusch, and Bogusch might well have appreciated the same of Raun when the latter's study began. If each had surveyed both plants and animals instead of only one or the other, we would already have some basis for an evaluation of historical changes in flora and fauna that might in turn reflect changes in the water relations or "hydrology" of the area. However, even had they done so, these two pioneers studied sites that were some distance apart, which would have been a complication for any comparisons between yesterday and today. We chose the previously unstudied invertebrates and the plants (in a forthcoming volume) for our own emphasis.

During the preparation of this book, a reviewer suggested that we describe how we went about our work. First, we consulted the scientific literature and topographic maps to see what was known about the flora and fauna and where the wetlands were located, respectively. Second, we obtained a Scientific Study Permit from Texas Parks and Wildlife Department. This is needed by anyone who collects plants and animals in state parks. Third, while making weekly trips to the public land of Palmetto State Park, where no permission to enter is required, we sent e-mails and letters and made phone calls to those who owned the private swamps and marshes surrounding the park. Even finding these names was a form of exploration, for we found it necessary to consult state and county officials, with almost no information to go on besides the location of the land itself. Most of the time we reached the right party, and whenever we did, permission was granted to enter, observe, and collect. We were pleasantly surprised by this courtesy, and the following cannot be emphasized too strongly: **Never trespass on private property.** It is the responsibility of everyone who takes to the field to be aware of his or her surroundings, not only the dangers presented by plants, animals, and the deep muck of the wetland itself but also the ownership of those very lands. The authors know of cases in which local landowners "escorted" trespassers off their property, and in similar scenarios far from these wetlands, reactions were much, much worse.

With verbal permission for private land and with a written scientific study permit for public land, we began photographing, collecting, and identifying as many plants and animals without backbones as we could. Invertebrates are much more difficult to identify than plants, birds, mammals, and reptiles, and for many species there are still no publications that allow nonspecialists to attach a scientific name to a specimen. When the library of the University of Texas at Austin did not have the relevant literature on its shelves, we attempted to obtain it from other institutions through Interlibrary Loan. When this failed, we sought out experts by e-mail, and in some cases we were forced to send the specimens in a box or vial. In short, and regrettably, with few exceptions reliable identification of insects, spiders, crustaceans, and other invertebrates cannot be made with field guides. It is a time-consuming and exacting task. As an aid in this process we provide measurements, usually under the "Remarks" section, for each species. These are body lengths of large specimens unless otherwise noted.

Invertebrates of Central Texas Wetlands

1
The Ottine Wetlands of Central Texas

The Ottine wetlands comprise a relict ecosystem that lies on the floodplain of the San Marcos River where it flows through Gonzales County in south-central Texas (Maps 1-1–1-3, Figs. 1-1, 1-2). They are named for their proximity to the village of Ottine and are unique for their geographic isolation from other Texas wetlands located farther east, along the coast, or in association with playa lakes in the Panhandle area. Notable too is their location astride the biogeographical line separating the eastern plants and animals of the United States from its western flora and fauna and for the resulting mixture of species that occurs there. The same may be said for the north-south direction, because this part of Texas is the northernmost limit for many plants and animals characterized as Neotropical elements, some of them occurring as far below the equator as Argentina. However, most animals and plants occurring here are eastern and Nearctic in their affinities.

According to the standard reference, Texas is divided into seven biotic zones or provinces (Blair 1950), and the Ottine wetlands are contained within the post oak woodlands of the namesake or "Texan" province. To the immediate east of this zone is the Austroriparian province, which represents an extension of the flora and fauna of the southeastern United States into Texas. Central Texas wetlands are more strongly affiliated with this zone than with any other. To the west of the Texan zone, and ironically even closer to the wetlands than the Austroriparian zone, lies the Balconian province identified with west Texas elements. To the south lies the Tamaulipan zone, which is home to Mexican and even Neotropical species that reach their northern limits here.

As of this writing there had been just over one-half century of research and publication dealing quite specifically with the Ottine wetlands. The earliest work was that of the botanist Bogusch (1928). Palmetto State Park was created soon afterward in 1933 (Shearer 1956; Maxwell 1970). Despite an

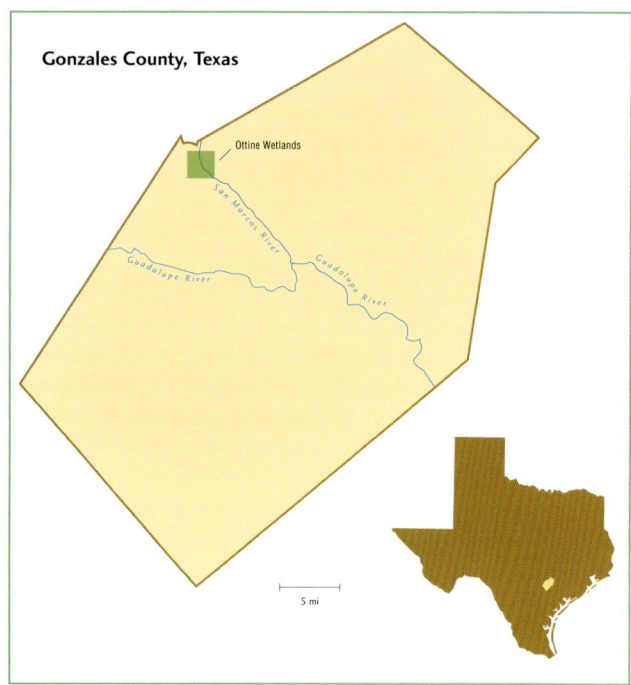

Map 1-1
The Ottine wetlands lie along the San Marcos River in Gonzales County, Texas (river width not to scale). Inset: The Ottine wetlands are located in south-central Texas. Precise location is at center of the square.

implied connection between the first scientific fieldwork and the first 198 acres acquired several years later for the protection of nearby flora and fauna, we found no evidence for any such link.

Early concern for the protection of these swamps and marshes may be explained by their unexpected occurrence so far west of their Austroriparian origin and by the resulting mix of eastern, western, and southern plants and animals that occur here and that are found in such associations nowhere else in the world. Research, reports, and efforts to excite public interest grew to include checklists of birds (Kirn 1935), amphibians and reptiles (Parks 1935b), and butterflies (Parks 1935c) while expanding upon the inaugural botanizing of Bogusch (Parks 1935a; Tharp 1935). Even plants that died long ago became objects of study. Cores taken from marshes and "bogs" suggest that the wetlands are at least twelve thousand years old and that their floral compositions have been changing with the drying climate ever since, though there is disagreement as to the extent of this change (Graham and Heimsch 1960; Patty 1968; Larson, Bryant, and Patty 1972). For example, some believe that birch, spruce, and fir trees once grew here, but if so, they have long since vanished. The same is true of sweetgum (*Liquidambar styraciflua*), a species that now occurs farther east among the botanically distinct wetlands of east Texas. For additional works dealing with plants of the Ottine wetlands' past, see

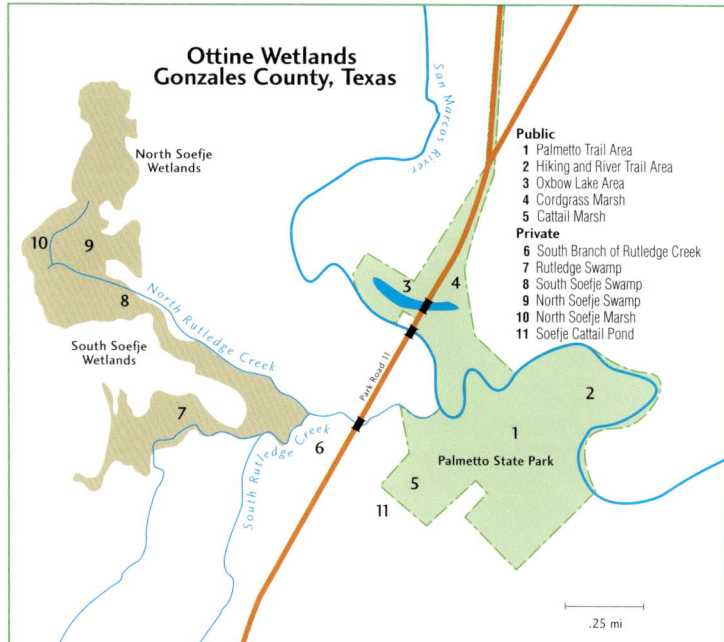

Map 1-2
The Ottine wetlands are both privately and publicly owned.

Ottine Wetlands
Gonzales County, Texas

North Soefje Wetlands

South Soefje Wetlands

San Marcos River

North Rutledge Creek

South Rutledge Creek

Park Road 11

Palmetto State Park

Public
1 Palmetto Trail Area
2 Hiking and River Trail Area
3 Oxbow Lake Area
4 Cordgrass Marsh
5 Cattail Marsh
Private
6 South Branch of Rutledge Creek
7 Rutledge Swamp
8 South Soefje Swamp
9 North Soefje Swamp
10 North Soefje Marsh
11 Soefje Cattail Pond

.25 mi

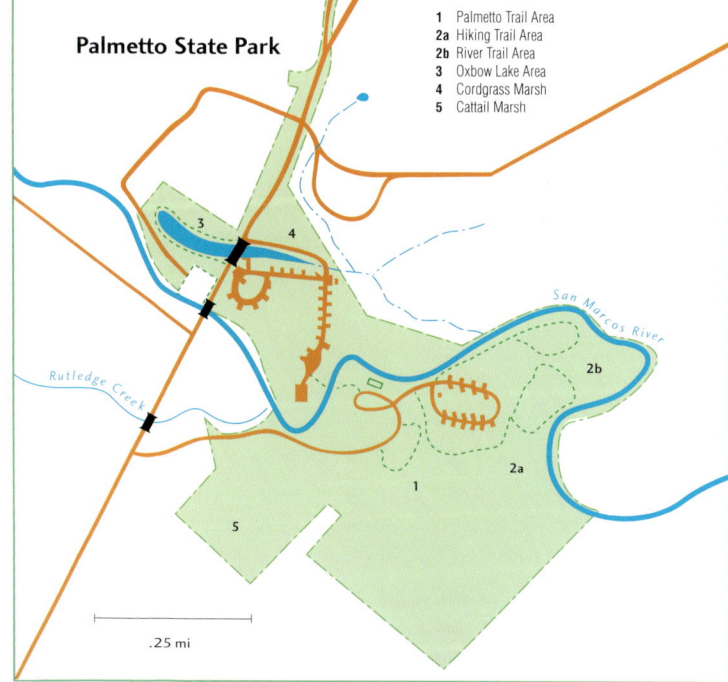

Map 1-3
Publicly owned wetlands lie within Palmetto State Park.

Palmetto State Park

1 Palmetto Trail Area
2a Hiking Trail Area
2b River Trail Area
3 Oxbow Lake Area
4 Cordgrass Marsh
5 Cattail Marsh

San Marcos River

Rutledge Creek

.25 mi

Fig. 1-1 The San Marcos River flowing near the entrance to Palmetto State Park

Chelf (1941), Plummer (1941, 1945), Graham (1958), Bryant (1977), and Mohlenbrock (2002). The flora of more eastern Texas wetlands may be compared to that of the central part of the state by consulting Rowell (1949), Nixon, Chambless, and Malloy (1973), and MacRoberts and MacRoberts (1998). For example, wetlands a little farther east support black gum (*Nyssa sylvatica*), and those nearly as far east as Louisiana support its close relative, tupelo gum (*Nyssa aquatica*). We saw neither species in the Ottine swamps and marshes, where the wetland flora and fauna of the southeastern United States reach their western limit.

The underlying geology of this portion of central Texas makes life possible for plants and animals that require abundant moisture and might not be expected so far west. Here rainwater percolates through porous sand derived from rock formations of Eocene age (roughly 50 million years old), reaches impenetrable layers below, and spreads laterally to regain the surface once more in the form of acidic seeps and springs. Particularly important is the Carrizo Formation. The geology and hydrology of this region have been detailed

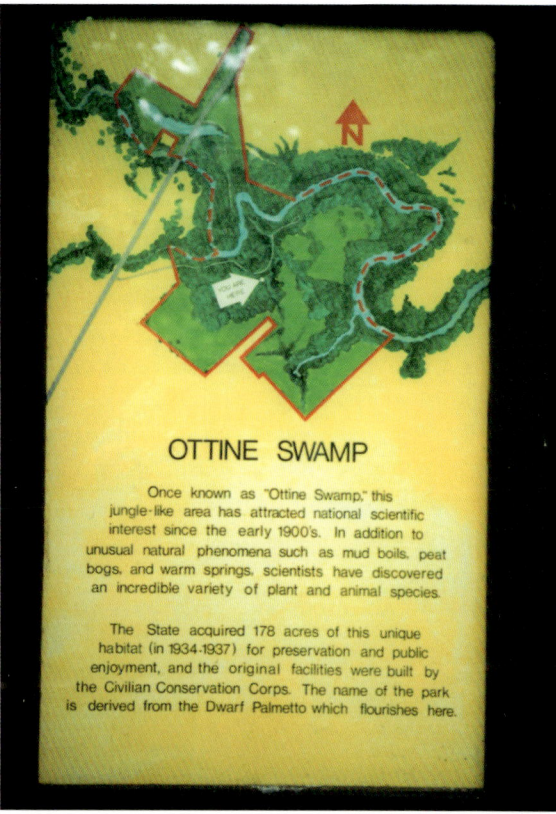

Fig. 1-2
Ottine Swamp as portrayed on the stone tower in Palmetto State Park.

by Cumley (1931), Bullard (1935), and King (1961). For a history of the area written at the time of Palmetto State Park's creation, see Hildebrand (1935). Recent bird checklists are those of Hartigan and Lasley (1987) and Rogers (1999). For the early work on mammals, reptiles, and amphibians, see Raun (1958, 1959).

To say that the Ottine wetlands remain poorly studied even after seventy-five years of work is an understatement. One landowner, while pointing us in the direction we sought, described our destination as "wilderness" that had not been visited for years. This oversight is remarkable considering the proximity of the site to several major universities that lie, in turn, near the center of a large, agriculturally oriented state. We knew or at least suspected the positions of some of the private wetlands because we had in hand the U.S. Geological Survey's topographical maps, the most recent of which was already twenty years old. Indispensable in this regard is the highly detailed map known as "Ottine Quadrangle, 7.5 minute series."

We were unable to examine two small areas that we wished to investigate despite several attempts to contact the landowners by letter. A lack of response is understandable, because the findings of biodiversity specialists often cause private citizens to regret their kindness at a later date. First among these sites is the plain beneath the overlook of Red Hill, lying just inside the northern entrance to Palmetto State Park. Its ash swamps and sphagnum marshes were the first of the Ottine wetlands to be studied (Bogusch 1928, 1930), in this case from an exclusively botanical viewpoint. Strangely, there is no indication that Bogusch was aware of the wetlands encompassed by the park that had not yet been established, although they were only a few miles south of his own site.

By the middle of the century the swamps and marshes studied by Bogusch had diminished drastically, were unrecognizable as such, and were perhaps eventually destroyed (Raun 1958). Thus, there might have been little for us to look at. Some of this change occurred within a decade of the botanist's field-work. At the time of our own visits we were unable to ascertain visually the situation even from the proximity of the Red Hill overlook because of intervening brush and trees. Tangles of trunks and denuded branches in winter were nearly as great an interference as the prolific leaves of spring and summer (Fig. 1-3).

Botanists had found sphagnum moss at this site, and its presence was remarkable as defining one western limit of sphagnum's range in the United States. Later reports of it came from the Soefje wetlands and from Hershop "bog" (Chelf 1941), the latter locality being the second of the two sites that we wished to explore but could not because we did not have permission to do so. That site is, or at least was, better described as a marsh than a bog and lies a short distance west of Palmetto State Park. Long ago the authority on this little peatland summed up its condition as "dead" and undergoing erosion and silting (Patty 1968). We devoted much time to the search for sphagnum on both private and public properties, but we never saw the three previously reported species (*Sphagnum imbricatum, S. palustre, S. subsecundum*) despite one report of success only thirty years before (Lodwick and Snider 1980). It appears that we walked over the same ground described in that publication.

When we encountered a wetland, we were faced with the task of determining its type or "species." This process has much in common with the identification of plants and animals, including the risk of misidentification, with its consequent ripple effects in future work undertaken by others (Vitt 1994). For example, it is tempting to classify some of the Ottine peatlands as "bogs," but a review of the literature convinced us that bogs in the strict sense of the word do not occur anywhere in south-central Texas despite the fact that others have used that term for some of the habitats we visited (Fig. 1-4). We settled on a classification that recognizes eleven wetlands of various types, five on public

Fig. 1-3 The view from Red Hill Overlook

Fig. 1-4 Peatland formerly described as a bog. The herbaceous plants in the center of the picture are soft rushes (*Juncus effusus*), and just behind are max myrtles (*Myrica cerifera*).

land and six on private land. The total area of all eleven combined lies some-where between one and two square miles.

Previous reports were available for some of our destinations. Most were old and found their original outlets in obscure publications or in unpublished the-ses. In one of two particularly valuable studies, more than seventy years had elapsed since the work was completed, and in the other, nearly half a century. Old literature such as this is sometimes more valuable than recent work because it allows comparisons of historical changes in habitats and more specifically in the compositions of their flora, fauna, and hydrology.

Wetlands are difficult to classify in a manner that satisfies everyone (Lewis 2001). In fact, the word *wetland* itself is subject to competing and complicated definitions. A good one is the straightforward version offered by the U.S. Fish and Wildlife Service, which sees wetlands as "lowlands covered with shallow and sometimes temporary or intermittent waters. They are referred to by such names as marshes, swamps, bogs, wet meadows, potholes, sloughs, and river-overflow lands. Shallow lakes and ponds, usually with emergent vegetation as a conspicuous feature, are included in the definition, but the permanent waters of streams, reservoirs, and deep lakes are not included. Neither are water areas that are so temporary as to have little or no effect on the develop-ment of moist-soil vegetation" (Mitsch and Gosselink 2000, 29). This concept emphasizes the presence of *plants* adapted to at least periodically saturated soil rather than the nature of the soil itself. This is one reason for our attention to botany as well as zoology. Yet the role of the master Ottine waterway, the San Marcos River, remains paradoxically uncertain in this interpretation except for those lands that become inundated by periodic floods.

All eleven Ottine wetlands are of the inland rather than coastal type, and most are probably relicts of a more humid age when eastern flora now in par-tial, if not full, retreat had spread west across central Texas from the eastern United States (Raun 1958). For example, a small patch of Gulf cordgrass (*Spartina spartinae*) grows in alkaline water that is at least slightly saline and is an ecological, if not historical, link to coastal marshy habitats located eighty miles to the southeast along the Gulf of Mexico. By some standards wetlands lying up to one hundred miles inland from the Gulf are considered con-stituents of the coastal plain community (Tiner 1999). Thus, the Ottine area is a living frontier with respect to sea and land as well as east and west and north and south.

We identified at least seven of the nine wetland types or "species" specifi-cally listed in the U.S. Fish and Wildlife Service definition. These seven are marshes, swamps, wet meadows, sloughs, river-overflow lands, shallow lakes, and shallow ponds (Mitsch and Gosselink 2000). It is tempting to exhaust the inventory by classifying some of the sites as potholes and bogs, but the Ottine area is too far south to embrace such features in the strict sense. For example, the curious bowl-shaped "lagoons" of Palmetto State Park could be described as potholes, but the latter are understood to be prairie features formed by glaciers that never extended as far south as Texas. Instead, the lagoons are believed to be natural depressions that were artificially modified during the construction of the park. Likewise, the true bogs of more boreal climes are fed exclusively by precipitation rather than by the additional sources of groundwater and surface water that are so vital to all of the Ottine wetlands (see Glossary for wetland definitions and distinctions). Sphagnum moss is also a bog indicator, though not as reliable as the precipitation criterion. Sphagnum once occurred here and perhaps still does. Neither we nor the Texas Parks and Wildlife Department know of any extant sphagnum in the Ottine wetlands (David Riskind, pers. comm., 2003). Eventually we settled on a streamlined wetland classification recognizing only four categories in the Ottine area: marsh, swamp, oxbow lake, and pond.

Finally, we note that wetlands may be permanent or intermittent, and they may be natural or artificial or a combination of both. Artificial or created varieties may be intentional reclamations from previously drained natural sites, they may be wetlands developed for the first time from drier uplands, or they may be accidental creations. The following list shows the eleven areas of the Ottine wetlands (Figs. 1-5–1-40).

Public Wetlands

1. Palmetto Trail area (swamp, natural/artificial, potentially permanent; Figs. 1-5–1-16)
2. Hiking and River Trail area (swamp, natural, intermittent; Figs. 1-17–1-19)
3. Oxbow Lake area (oxbow lake, natural/artificial, potentially permanent; Figs. 1-20–1-21)
4. Cordgrass Marsh (marsh, natural/artificial, potentially permanent; Figs. 1-22–1-23)
5. Cattail Marsh (marsh, artificial, potentially permanent; Fig. 1-24)

Private Wetlands

6. South Branch of Rutledge Creek (swamp, natural, permanent; Figs. 1-25–1-28)
7. Rutledge Swamp (swamp, natural, permanent; Figs. 1-29–1-33)
8. South Soefje Swamp (swamp, natural, permanent; Figs. 1-34–1-35)
9. North Soefje Swamp (swamp, natural, permanent; Fig. 1-36)
10. North Soefje Marsh (marsh, natural, permanent; Figs. 1-37–1-39)
11. Soefje Cattail Pond (pond, artificial, potentially permanent; Fig. 1-40)

Five of these wetlands are public, and six are private. There are six swamps, three marshes, one oxbow lake, and one pond. Six wetlands are entirely natural in their water sources, but the two cattail regions are artificial in both origin and maintenance. The three remaining wetlands exist through combinations of nature and human intervention via wells and/or pipelines installed or maintained by state or private agencies.

Water source is also relevant to classification by the criterion of permanence. Some natural wetlands remain wet year-round and give all indications of remaining in this condition for the foreseeable future without human intervention. These we classify as permanent, and there are five of them. Potentially permanent wetlands are those that would probably dry out at least during the summer months if not for at least one artificial source of water that can be called upon to prevent desiccation. There are five of these. If a wetland lacks both an artificial source and a permanent natural source of water, it will eventually dry out unless and until it is fed by rains and/or flooding. We classify this condition as intermittent. There is only one such area on our list, the publicly owned Hiking and River Trail region of Palmetto State Park, which the flooding San Marcos River occasionally transforms into an ephemeral swamp. We distinguish this spot from the adjacent Palmetto Trail region because it does not have the benefit of the latter's artesian well. A few permanent or nearly permanent ponds and similar bodies of water did not warrant separate status in our list because of their small size.

Overview of the Ottine Wetlands

1. Palmetto Trail Area (Figs. 1-5–1-16)

The namesake species of Palmetto State Park, the dwarf palmetto (*Sabal minor*), may be seen here in abundance bordering a luxurious, winding path that skirts lagoons and green ash trees. Part of this wetland is supplied by an artesian well. The water, delivered by the agency of an old ram pump and by nearly negligible dripping runoff from a wood and stone tower, has a neutral pH of 7.0 and a temperature of 71°F. Most of the Palmetto Trail area must rely upon seasonal rains and the flooding San Marcos River to fill the numerous lagoons that are its main features.

In summer the vertical pipe of the artesian well is likely to cease supplying its pulsating fountain of water. Yet some flow usually continues through the ground-level horizontal pump that makes water more reliable here than elsewhere.

This Palmetto Trail area, as small as it is, offers more to the casual visitor per unit time both day and night than any other area accessible to the public.

Fig. 1-5
Dwarf palmetto
(*Sabal minor*)

Fig. 1-6
Lagoon in the
Palmetto Trail
area (wetland 1)

Fig. 1-7
Green ash (*Fraxinus penn-sylvanica*) strangled by
enormous Alabama supple-jack vine; dwarf palmetto at
lower left (wetland 1)

Fig. 1-8
Artesian well (right foreground)
and water storage tower (wetland 1)

Fig. 1-9
Hydraulic ram pump water-
ing lagoon (wetland 1)

Fig. 1-10
Stone roundhouse with
statement of the pal-
metto's changing geo-
graphic distribution
(wetland 1)

Fig. 1-11
Mexican palm
(*Sabal mexicana*;
wetland 1)

Fig. 1-12
Yellow flag (*Iris
pseudacorus*;
wetland 1)

Fig. 1-13
Purple fleur-de-lis
(*Iris hexagona*;
wetland 1)

Fig. 1-15
Nine-banded armadillo (*Dasypus novemcinctus*) foraging at night (wetland 1).

Fig. 1-14
Black vultures (*Coragyps atratus*) coming to roost in winter (wetland 1).

Fig. 1-16
Lagoon drying out in summer (wetland 1).

Trees growing here include green ash (*Fraxinus pennsylvanica*), boxelder (*Acer negundo*), hackberry (*Celtis laevigata*), bur oak (*Quercus macrocarpa*), shumard oak (*Q. shumardii*), osage-orange (*Maclura pomifera*), cedar elm (*Ulmus crassifolia*), winged-elm (*U. alata*), anaqua (*Ehretia anacua*), and others mentioned in the following pages.

Not to be confused with the smaller native dwarf palmetto is the cabbage palm (*Sabal palmetto*) of the eastern United States or the Mexican palm (*S. mexicana*) of extreme southern Texas. During our studies one tall, suspicious-looking specimen distinguished by a massive trunk towered above a sign near the park entrance that encouraged protection of plant life. Its fruits and flowers were not available, so we fell back on microscopic leaf structure as a means to identify the individual to species level (Zona 1990). According to that method of identification the plant is an exotic Mexican palm rather than a native dwarf palmetto, with origin unknown. The history of the Mexican palm in Texas was being revised at the time of writing (Lockett 2003). Other notable shrubs in addition to the dwarf palmetto include buttonbush (*Cephalanthus occidentalis*), red buckeye (*Aesculus pavia*), and rough-leaved dogwood (*Cornus drummondii*).

A remarkable forb growing along the trails is Florida lettuce (*Lactuca floridana*). In spring and summer it grows to heights exceeding twelve feet. By this time the ubiquitous swamp katydid (*Amblycorypha oblongifolia*) has molted from juvenile to adult, and a hundred or more individuals of both sexes crowd onto a single plant, where they defoliate the leafy giant until little more than the central stem remains.

The most popular botanical attractions in the public wetland are two iris species. Both may be seen with good timing and good luck on or near the Palmetto Trail. The introduced yellow-flowered species known as yellow flag (*Iris pseudacorus*) is a more reliable sight that seems to be pushing aside the smaller, purple-flowered native species known as purple fleur-de-lis (*Iris hexagona*). We found no records of the exotic species in the literature, but there are less-threatened populations of the native growing in swamps on private lands nearby. Occasionally someone confuses the abundant giant spiderwort (*Tradescantia gigantea*) with the less prolific native iris because both bloom early in spring and both bloom in one shade of blue or another.

Other plants along the Palmetto Trail include the red-fruited Carolina wolfberry (*Lycium carolinianum*), white-flowered frostweed (*Verbesina virginica*), poison ivy (*Toxicodendron radicans*), Virginia creeper (*Parthenocissus quinquefolia*), and Alabama supplejack (*Berchemia scandens*).

Beginning in late afternoon one may hear a racket of flapping wings in the vicinity of the water tower. These, according to our experience, are black vultures (*Coragyps atratus*), congregating to roost for the night. Hissing, croaking, and the sound of blows delivered by jostling forelimbs fill the air as the big birds compete for perching space high up in the trees. Turkey vultures (*Cathartes aura*) may roost here too. Hikers who walk the trail after dark should be prepared for sounds that are disturbing even when one is sure of their source. The nine-banded armadillo (*Dasypus novemcinctus*) makes loud crashing noises on the ground as it forages for insects after dark. This immigrant from Mexico seems oblivious to quiet observers and will sometimes walk right up to a pair of boots.

During our studies all of the lagoons dried up during the summer, though the oxbow lake of wetland 3 did not. Raun (1958) noted precisely the opposite condition, reporting that the lagoons never went dry even when the oxbow lake was empty. He attributed their longevity to the overflow from a "pump house" and water from a "spring." As far as we can tell, this must refer to the water storage tower and to the artesian well, respectively, neither of which was up to the job of keeping even one lagoon wet in the summer of 2001. Raun also seemed to report introduced water hyacinths that were not present when we arrived. There might be some confusion because he identified them by the scientific name of one of the two iris species that grow among the lagoons.

2. Hiking and River Trail Area (Figs. 1-17–1-19)

With the exception of a few local seeps feeding small ponds of very limited extent, such as that of the extinct mud boil, there is little reliable water here; although there is a creek that has cut a small canyonlike gorge where a visitor leaving the trail could take a dangerous tumble. The wetland is often nearly dry, at least on the surface.

Closest to the Palmetto Trail is the Hiking Trail; farther beyond is the River Trail. Many of the wetland trees and shrubs treated by us in a forthcoming companion volume may be seen while walking the loop, but some species, including black willow (*Salix nigra*), bald cypress (*Taxodium distichum*), Texas persimmon (*Diospyros texana*), cottonwood (*Populus deltoides*), pecan (*Carya illinoinensis*), mulberry (*Morus rubra*), osage-orange (*Maclura pomifera*), sassafras (*Sassafras albidum*), wax myrtle (*Myrica cerifera*), Roosevelt weed (*Baccharis neglecta*), and elderberry (*Sambucus canadensis*), are more conspicuous elsewhere in the park and on private land.

3. Oxbow Lake (Figs. 1-20–1-21)

This small, shallow lake was formed long ago when the San Marcos River changed course and left standing water behind. In fact, its name is merely the term used to designate all such relict riverbeds. Cut off from the San Marcos, it relies for its water upon rain, temporary reunion with the river via floods, and two artificial sources. One of these is an artesian well, and at the time of

Fig. 1-17 Former site of mud boils, now a pond colonized by grass-leaf arrowhead (*Sagittaria graminea*; wetland 2)

Fig. 1-18
Creek cutting deeply through the soil on its way to the San Marcos River in the Hiking and River Trail area (wetland 2)

Fig. 1-19
Black willow (*Salix nigra*) with shelf fungi

Fig. 1-20
The oxbow lake,
viewed from its west
end (wetland 3)

Fig. 1-21
Artesian well
near oxbow lake
(wetland 3)

writing the other was a pipeline originating on the property of the Warm Springs Foundation (Park Superintendent Mark Abolafia-Rosenzweig, pers. comm., 2001). The pH of the oxbow lake is 7.0, or neutral, and the pH of the nearby artesian well is slightly higher at 7.1. The oxbow lake is said to be no more than six feet deep in most places, and though it appeared to be stable during our study, it has gone dry in past decades for three years at a time (Raun 1958). Along its shores grow the tall southern reed (*Phragmites australis*), native bristlegrass (*Setaria scheelei*), and bald cypress (*Taxodium distichum*).

Fig. 1-22
Gulf cord-
grass marsh
with Malaise
trap in place
(wetland 4)

Fig. 1-23
Old but still bub-
bling geothermal
well near cord-
grass marsh
(wetland 4)

4. *Cordgrass Marsh* (Figs. 1-22–1-23)

In 2001 the Gulf cordgrass marsh could be supplied at will with water by a pipeline that stretched across the wetland from one side to the other on its way to the oxbow lake, which can be supplemented by the same pipe. The source on the adjacent Warm Springs Foundation property presumably delivered water from a well with a temperature of 71°F, though an ancient geothermal well still bubbled when we visited the grounds. Its temperature was 98.1°F and thus only slightly lower than human body temperature. Rains and the San

Marcos River in flood are additional sources of water for the cordgrass marsh. The pH of the water here is unique for its alkalinity. Far from the acid condition expected of "boggy" areas, it was slightly higher than 8.0, and by far the highest pH we discovered in these wetlands.

This somewhat isolated marsh has been estimated to be ten acres in extent (Raun 1959). It is located in a shallow basin just southeast of park headquarters as they stood at the time of writing and is easily missed from the road. The dominant plant is Gulf cordgrass (*Spartina spartinae*), also known as sacahuiste, which grows in characteristic clumps in shallow and at least slightly saline water. Here too is the nonnative, weedy cattail (*Typha domingensis*), which seems to have largely, if not entirely, replaced the native broad-leaved species (*T. latifolia*) recorded by all those who studied the area before us (Bogusch 1930; Raun 1958; Parks 1935a). Growing nearby is a stand of southern reed (*Phragmites australis*), a tall true grass that grows abundantly along the margin of the adjacent oxbow lake. It extends into the cordgrass patch but is uncommon elsewhere. Other plants include Carolina wolfberry (*Lycium carolinianum*), Roosevelt weed (*Baccharis neglecta*), mesquite (*Prosopis glandulosa*), downy hawthorn (*Crataegus mollis*), and switchgrass (*Panicum virgatum*). Most of these may be seen as colonizers or invaders of the marsh.

In the mid–twentieth century this tiny wetland dried out so severely that the clay soil cracked (Raun 1958). We don't know the fate of a second cordgrass meadow, studied by Bogusch in the early part of the century. It was located on private land near the northern entrance to Palmetto State Park. Perhaps it has vanished completely.

5. Cattail Marsh (Fig. 1-24)

After familiarizing ourselves with maps and the results of previous studies, we were surprised to find a cattail marsh in the hinterland of Palmetto State Park. Perhaps it was recently formed. There is certainly no mention of this wetland in a previous survey (Raun 1958). It relies for its water on runoff from an artesian well that gushes even in summer on nearby private land. This is the same well that supplies the cattail pond on that property, and we believe that these plants founded the marsh through their progeny. The pH here is 7.1.

Despite the force of the well's flow the water evaporates and sinks into the ground during the hottest months before it reaches this clearing. The soil then becomes cracked and dry, and thus, at least at the surface, the water in the small artificial wetland is seen to be ephemeral or temporary. It is unclear if the marsh would survive if the well on private property should stop flowing. Perhaps both its maintenance and its origin can be traced to human disturbance that can be viewed, as it almost never is, in a favorable light.

Fig. 1-24 Cattail marsh with Malaise trap in place (wetland 5)

The small patch is less than one acre in extent. It is not on any trail and can be reached only by a short hike into the woods. We searched high and low in the most literal sense but found no evidence of previous flowering by the cattails growing here. Thus, positive identification was not possible. We believe them to be the exotic giant cattail (*Typha domingensis*), and if this is so, then the native broad-leaved cattail (*T. latifolia*), the only species recorded previously (Bogusch 1928; Raun 1958), would appear to be extirpated from the Ottine wetlands, because we did not find it at any of the sites where cattails now occur.

In this hinterland cattails flourish alongside sallow caric sedge (*Carex lurida*), which is green in winter when the former plants have died back aboveground to dry brown leaves and stems. Here also is the halberd-leaved hibiscus (*Hibiscus laevis*). Together the low-growing species survive in a clearing surrounded by towering black willows (*Salix nigra*), boxelder (*Acer negundo*), and green ash (*Fraxinus pennsylvanica*).

6. South Branch of Rutledge Creek (Figs. 1-25–1-28)

This, the first of the private wetlands treated in detail here, is a complex array of permanent ponds, seeps, shaded swampy patches, a pecan-rich floodplain that receives water from the San Marcos River, and Rutledge Creek that

Fig. 1-25
Floodplain pecan grove
(wetland 6)

Fig. 1-26
At the center of the Ottine
wetlands; fork of Rutledge
Creek where its north and
south branches merge on
the way to the San Marcos
River (wetland 6)

Fig. 1-27
Rutledge Creek
with grape lianas
(wetland 6)

Fig. 1-28
Feral hog skull
(wetland 6)

empties into the same. Like all but one of the private lands its waters arise from natural sources only. We measured a pH of 6.0 in Rutledge Creek, so it may be described as acidic. In such peatlands as these one sometimes experiences the phenomenon known as "quaking." Quaking occurs when the ground beneath the feet quivers like jelly with each step, caused by a thick growth of vegetation spreading out across the top of the wetland's water.

Much of the land endures periodic heavy flooding from the San Marcos River, and as a result pecan trees flourish in tall stands as they do nowhere else. Trees and shrubs are like those of the public trails across the highway except that black willow, cottonwood, elderberry, and, of course, pecan are more abundant. Feral hogs moved into the area several decades before our study and are now both common and dangerous (Taylor 1991).

7. *Rutledge Swamp* (Figs. 1-29–1-33)

Rutledge Swamp is the best example in the Ottine area of a permanent swampland more typical of the southeastern United States. It is supplied by seeps from underlying groundwater in the Carrizo Formation of Eocene age. These yielded a pH of 5.5, one of the lowest or most acidic values measured in the Ottine area, and thus quite opposite in nature from the alkaline waters of the cordgrass marsh. Water temperature was typically and predictably 71°F. At the time of our study the muck was treacherously deep, and the going was difficult. Water is liable to enter the tops of sinking knee-high boots with unpleasant consequences, and the extraction of rapidly flooding footwear may prove difficult. The foot itself might tear loose from a boot that remains mired in muck. It is a good idea to travel in pairs in such a wetland.

Within the surrounding forested lowland that lies between pasture and swamp, a spectacular windthrow marked the approach to our goal. Among the trees were giant cottonwoods and shumard oaks. These did not fall because of high winds per se but because of unusually weak root systems that did not penetrate deeply enough into the oxygen-poor soil to provide solid anchors. Not so susceptible to windthrow are the ashes, boxelder, willow, hackberry, and pecan. Dwarf palmettos are abundant here, as are wax myrtle (*Myrica cerifera*) and the occasional scythe-fruit arrowhead (*Sagittaria lancifolia*).

Deadly water hemlock (*Cicuta maculata*) is so poorly anchored in the muck that a boot splashing down near the plant may cause it to lean precariously. If an unsuspecting human should happen to chew on its leaves, stems, or roots, the experience may prove fatal. Dangerous animals that may be encountered here include canebrake rattlesnakes and western cottonmouths.

In a few clearings huge cinnamon ferns (*Osmunda cinnamomea*) grow to heights of six feet in the company of wax myrtle and yaupon that flourish on drier ground nearby. One expects a dinosaur to peer through such foliage. Cat-

Fig. 1-29
Seep flowing through Rutledge Swamp (wetland 7)

Fig. 1-30
Wind-thrown shumard oak (*Quercus shumardii*); wetland 7)

Fig. 1-31
Canebrake rattlesnake (*Crotalus horridus atricaudatus*) digesting recent meal beneath fallen willow log (*Salix nigra*)

Fig. 1-32
Juvenile canebrake rattlesnake (*Crotalus horridus atricaudatus*) resting several feet aboveground

Fig. 1-33
Juvenile western cottonmouth (*Agkistrodon piscivorus leucostoma*) preparing to strike

tails were once reported in Rutledge Swamp (Raun 1958), but we saw none in our own explorations. Perhaps these early colonizers have been replaced by other plants.

During the mid-twentieth-century drought Rutledge Swamp dried up (Raun 1958). In conversation with one of its current owners who was aware of that history, we learned that it had not done so during the past several decades.

Fig. 1-34
Photographing cocklebur in clearing (wetland 8)

Fig. 1-35
Sassafras grove on upland border (wetland 8)

8. *South Soefje Swamp* (Figs. 1-34–1-35)

South Soefje Swamp is supplied with seeps from the same rock formation that supplies the more northern Soefje wetlands and Rutledge Swamp. Green ash and boxelder grow abundantly, and in a small, open, marshy area, balloon-vine (*Cardiospermum halicacabum*) and cocklebur (*Xanthium strumarium*) dominate the scene. Peat has accumulated over thousands of years to depths in some areas exceeding fifteen feet (Graham 1958). Along a slope that is the interface between upland pasture and the swamp we were surprised to find a small sassafras (*Sassafras albidum*) grove that may well lie on the western edge of this species' range.

Fig. 1-36 Ash swamp (*Fraxinus pennsylvanica*; wetland 9)

9. North Soefje Swamp (Fig. 1-36)

Like South Soefje Swamp, this wetland receives a permanent or nearly permanent supply of cool water from numerous seeps in the underlying sand. We found the water more acidic than elsewhere, with a pH as low as 5.0. It flowed from seeps at temperatures ranging from 66°F to 72°F. Some of this variation might be due to shaded versus more exposed conditions because a few seeps lie at the base of a wax myrtle or some other shrub or tree. On the western edge of North Soefje Swamp water accumulates to form ponded, grass-crowded avenues that suggest a Louisiana bayou. Green ash is the defining swampland tree here. It flourishes alongside black willows and boxelder.

Those forested peatlands that have been described in the past as bogs lie in this area and in the nearby South Soefje wetland. In reality there are no bogs anywhere in the Ottine wetlands because none of these peatlands rely upon precipitation as their sole source of water. Quite to the contrary, at present hillside seeps seem to constantly moisten exposed clearings of black peat surrounded by trees and shrubs. Long before our studies began, consideration was given to the possibility of mining these sites commercially (Chelf 1941; Plummer 1941, 1945).

According to Raun (1958, 1959) this swamp and the adjacent marsh described in the next section dried up during the devastating seven-year drought of the mid–twentieth century. According to the evidence of pollen samples (Graham 1958; Bryant 1977), these wetlands have survived much as they are now for at least the last eight thousand years, despite presumably recurrent catastrophes.

10. North Soefje Marsh (Figs. 1-37–1-39)

In this wetland the same seeps that supply the adjacent tree-dominated swamp maintain an open marsh of primarily herbaceous vegetation. More conspicuous here than elsewhere is giant cutgrass (*Zizaniopsis miliacea*), which must often be mistaken for cattail. The latter does not grow at this site as far as we could tell. If a visitor jumps up and down on thick sedge near the seeps, the quaking phenomenon may be experienced, as described earlier.

Alongside giant cutgrass grows sallow caric sedge (*Carex lurida*), southern wax myrtle (*Myrica cerifera*), dwarf palmetto (*Sabal minor*), coffee senna (*Senna*

Fig. 1-37 Cool, acidic seep from Carrizo Formation (wetland 10) with overhanging vines of Alabama supplejack (*Berchemia scandens*)

Fig. 1-38
Working through giant cutgrass (*Zizaniopsis miliacea*) in winter (wetland 10)

Fig. 1-39
Giant cutgrass marsh with Drummond rattlebox (*Sesbania drummondii*) shrubs (wetland 10)

occidentalis), halberd-leaved hibiscus (*Hibiscus laevis*), and balloon-vine (*Cardiospermum halicacabum*). This wetland was the largest marsh we saw in the Ottine area and one of only a very few of any size.

Fig. 1-40 Giant cattail pond (wetland 11)

11. Soefje Cattail Pond (Fig. 1-40)

The cattail pond is an artificial wetland maintained at the time of writing by a gushing artesian well. Water seeping through the banks of the pond also sustains the nearby publicly owned cattail marsh of Palmetto State Park. Long ago a fish hatchery stood on the spot and was hailed as the most reliable refuge for aquatic life in the midst of droughts that dried up nearly every other source of water (Raun 1958). This insurance is no longer extant.

2
Dragonflies and Damselflies

Mammals, reptiles, amphibians, and to a lesser extent the fishes of the Ottine wetlands were studied long ago (Raun 1958, 1959). Birds were treated before and after those surveys by extensive checklists (Kirn 1935; Hartigan and Lasley 1987; Rogers 1999). Here we emphasize the remaining task, the diversity of wetland invertebrates, and we begin with the dragonflies that so appropriately begin their own lives underwater.

Comanche skimmer
Libellula comanche
(FIG. 2-1)

Biology: We saw this attractive blue dragonfly at a single seepage near an artificial pond. It is known to favor springs and slow sections of streams

Fig. 2-1
Comanche skimmer (*Libellula comanche*)

and rivers. Males patrol slowly and perch on tall weeds (Dunkle 2000) and, as we discovered, on the tips of tree branches overhanging the seep.

Distribution: Comanche skimmers are among the few western animals of a wetland that is generally regarded as a relict of an eastern encroachment. They occur from the Pacific Ocean to an eastern limit near the Ottine swamps.

Remarks: Length = 55 mm. Blue body, white face, and white spots near the wing tips provide a unique and unmistakable combination.

Similar species: None.

Fig. 2-2
Great blue skimmer (*Libellula vibrans*)

Great blue skimmer
Libellula vibrans
(FIG. 2-2)

Biology: Great blue skimmers are large and rather fearless dragonflies of swamps and marshes. Elsewhere they are known to frequent pools and slow forest streams. Our experience supports the observation of one authority: "This regal species is often tame, allowing a close approach as it perches on a shaded twig." (Dunkle 2000, 183).

Distribution: From the Atlantic Ocean to a western limit near the Ottine wetlands.

Remarks: Length = 63 mm. This is the largest skimmer in the United States and easier to bag with a net than most. We never saw it or the Comanche species in the nearby semiarid uplands of the Lost Pines forest.

Similar species: The great blue skimmer is larger than the slaty skimmer and lacks the white face of the Comanche skimmer.

Fig. 2-3
Widow skimmer (*Libellula luctuosa*)

Widow skimmer
Libellula luctuosa
(FIG. 2-3)

Biology: Widow skimmers range far from water much as the yellow-sided skimmers do. Males seldom guard females while they lay eggs, and when transforming to the winged adult stage, the nymphs or larvae crawl ashore to cast off their skins on grass near the pond (Needham and Westfall 1954).
Distribution: From coast to coast.
Remarks: Length = 50 mm. While resting in clearings, the widow can be approached quite closely before it takes flight.
Similar species: None.

White-tailed skimmer
Libellula lydia
(FIG. 2-4)

Biology: Males are unmistakable for their white abdomen and black-banded wings. They patrol ponds and swamps in search of prey and mates and in defense of territories, which they hold against competing males. They threaten each other by raising the distinctively colored abdomen like a flag (Dunkle 1989). Female white-tails are darker in color, are not as conspicuous as the males, and are so unlike their mates in appearance that they are likely to be mistaken for a different species altogether. Mating occurs on the wing, and over time the female deposits about one thousand eggs by dipping the tip of her abdomen at the surface of the pond as she hov-

Fig. 2-4
White-tailed
skimmer
(*Libellula
lydia*)

ers above it. Her mate guards her against other males during the process
but is helpless to prevent small fishes from gulping down the eggs. We
observed this on several occasions, and it appears that fishes are attracted
to the surface disturbances made by egg-laying females. Both sexes are
sometimes found far from water (Walker and Corbet 1975).

Distribution: From coast to coast within the United States.

Remarks: Length = 48 mm. It is said that males will sometimes perch on
bystanders. We never experienced this and, quite to the contrary, found
all dragonflies unapproachable during daylight hours beyond capture with
a net. The twelve-spotted skimmer (*L. pulchella*) occurs in all forty-eight
contiguous states and often flies alongside the white-tailed skimmer
(Needham and May 2000). We did not see the twelve-spot in the wetlands. Males have much more gaudily patterned wings, but females of the
two species are easily confused.

Similar species: None.

Slaty skimmer
Libellula incesta
(FIG. 2-5)

Biology: This species prefers muddy-bottomed ponds for the development of
its young. Males are a dark, slaty blue and are most active in the morning.
They guard their mates while the females deposit eggs in the water (Dunkle 1989).

Distribution: From the Atlantic Ocean to a western limit somewhere in
Texas.

Fig. 2-5
Slaty skimmer
(*Libellula
incesta*)

Remarks: Length = 52 mm. This species is not as abundant or at least not as apparent in the swamps as it is in the nearby uplands of the Lost Pines.
Similar species: None.

Fig. 2-6
Roseate skimmer (*Orthemis ferruginea*)

Roseate skimmer
Orthemis ferruginea
(FIG. 2-6)

Biology: Brightly colored males are territorial and fight one another until a single resident dominates a pond. Successful individuals fly from a favorite perch to battle competitors, and after mating with a female, they guard her from other males. Mated females deposit their eggs by splashing the surface of the pond with the tip of the abdomen. This hurls the eggs onto the shore just above the waterline or onto emergent vegetation (Harvey and Hubbard 1987).

Distribution: In the United States the roseate skimmer occurs from coast to coast.

Remarks: Length = 55 mm. The male's red to magenta color makes this species more visible from a distance than any other species except the brilliant neon skimmer.

Similar species: None.

Fig. 2-7
Neon skimmer
(*Libellula croceipennis*)

Neon skimmer
Libellula croceipennis
(FIG. 2-7)

Biology: The male is a beautiful red dragonfly, and the female is a drabber brownish color. We saw more individuals of both sexes here than we did in the nearby uplands of the Lost Pines forest. Males are especially conspicuous for their habit of perching atop tall, dead stems in open areas.

Distribution: In the United States the neon skimmer occurs from eastern Texas to California with a gap in southern New Mexico.

Remarks: Length = 57 mm. Striking color makes this the most beautiful of the Ottine dragonflies. According to one authority, the male actually seems to glow while perched (Dunkle 2000). It retains its beauty in the collection box.

Similar species: A second red species, *L. saturata,* might also occur in the uplands. It is more thoroughly red than the species treated here, including the veins of the wings.

Fig. 2-8
Yellow-sided skimmer (*Libellula flavida*)

Yellow-sided skimmer
Libellula flavida
(FIG. 2-8)

Biology: Both sexes will forage in clearings far from the nearest body of water. The behavior of the yellow-sided skimmer must be similar to that of the other members of its genus, but there appears to be very little published on the subject. Its preferred habitats are said to be mucky or boggy spring seepages (Dunkle 2000), and thus we find it remarkable that the species is not more common here.

Distribution: From the Atlantic Ocean to a western limit in Texas just east of the Pecos River.

Remarks: Length = 51 mm. The yellow-sided skimmer provides another example of the great differences between the sexes of some dragonflies. The blue male could easily be mistaken for a species different from that of the female.

Similar species: None.

Fig. 2-9
Hyacinth glider (*Miathyria marcella*)

Hyacinth glider
Miathyria marcella
(FIG. 2-9)

Biology: Larvae of this species live in quiet water among the hanging roots of exotic water hyacinth, and for this penchant the dragonfly received its common name. We noticed the introduced plant in a single wetland, a private pond near Palmetto State Park, and our encounters with the adult insect were always nearby. Male hyacinth gliders patrol above the water hyacinth mats and mate with females in flight. The female lays her eggs among the plants while the two remain coupled (Dunkle 2000).

Distribution: This animal is not native to Texas or to the United States. It was introduced long ago from South America with the aquatic weeds favored by their larvae. Its current range in the United States is said to be extremely limited, confined to a thin strip of the Gulf Coast region from Florida to Brownsville, Texas, with occasional strays wandering inland. However, hyacinth gliders appear to be established in the Ottine wetlands.

Remarks: Length = 40 mm. Hyacinth gliders were probably more common in the wetlands prior to our study, because the lagoons of Palmetto State Park were reportedly once crowded with water hyacinths. The plants were removed in deference to native flora.

Similar species: None.

Fig. 2-10
Spot-winged glider (*Pantala hymenaea*)

Spot-winged glider
Pantala hymenaea
(FIG. 2-10)

Biology: Spot-winged gliders are primarily denizens of temporary rather than permanent bodies of water. They are more erratic in flight than their close relative, the wandering glider, and less prone to hover (Dunkle 2000).

Distribution: From coast to coast within the United States.

Remarks: Length = 50 mm. This dragonfly migrates in swarms along the Atlantic Coast and turns up in deserts fifty miles from the nearest body of water (Dunkle 2000).

Similar species: The wandering glider lacks the dark spot on the hind wing that characterizes the spot-winged glider.

Wandering glider
Pantala flavescens
(FIG. 2-11)

Biology: The wandering glider breeds in rain pools and sometimes in slowly moving water. We saw few specimens, and these were generally spied in open areas near ponds and seeps. Males patrol territories and sometimes remain coupled to females while the latter lay their eggs. On hot days the abdomen is allowed to droop, thus minimizing the exposure of the body to the sun.

Distribution: Coast to coast in the United States.

Remarks: Length = 50 mm. This is the world's most widely distributed dragonfly and has been caught on ships at sea (Needham and May 2000).

Fig. 2-11
Wandering glider (*Pantala flavescens*)

Perhaps for this reason and because it flies both day and night, it has also been described as "the world's most evolved dragonfly" (Dunkle 2000, 216).

Similar species: See spot-winged glider.

Illinois river cruiser
Macromia illinoiensis
(FIG. 2-12)

Biology: As the common name suggests, this dragonfly has a preference for flowing waters, including those of rivers and streams. They do forage at great distances from these haunts, and we encountered our few specimens

Fig. 2-12
Illinois river cruiser (*Macromia illinoiensis*)

in wooded areas closer to ponds than to running waters. In the northern half of the United States, breeding also occurs in lakes, where larvae seem to prefer living among the green algae known as stoneworts (*Chara* spp.; Needham and May 2000). Males patrol territories along the riverbank, mostly in the morning, and mating pairs sometimes hang from trees (Dunkle 2000).

Distribution: From the Atlantic Ocean to a western limit in south-central Texas.

Remarks: Length = 76 mm. River cruisers take advantage of the regularity of cornfields by flying up and down the rows in search of insect prey (Dunkle 2000).

Similar species: The stream cruiser resembles the river cruiser but is more brown than black and is about 15 mm shorter.

Fig. 2-13
Stream cruiser
(*Didymops transversa*)

Stream cruiser
Didymops transversa
(FIG. 2-13)

Biology: Stream cruisers are said to be among the first dragonflies to appear in spring, and our experience corroborates this, for it was the first species caught in our net (mid-March). Their preference for flowing waters was also borne out—our specimen was netted while it patrolled up and down the banks of the San Marcos River.

Distribution: From the Atlantic Ocean to the Pecos River of Texas.

Remarks: Length = 60 mm. While foraging away from water, this dragonfly is more likely to hover than is the somewhat similar Illinois river cruiser (Dunkle 2000).

Similar species: See Illinois river cruiser.

Fig. 2-14
Black saddle-
bags (*Tramea
lacerata*)

Black saddlebags
Tramea lacerata
(FIG. 2-14)

Biology: We encountered the black saddlebags as it foraged at the edge of a pasture adjacent to Soefje Swamp. The common name derives from the shape and position of black markings on the dragonfly's wings. It forages rather high above the ground and has been described as an "exceedingly agile flyer" (Dunkle 2000, 218), consistent with our own discovery that it is harder to catch than most other species. Males patrol their pond territories at speeds of up to seventeen miles per hour. Females often lay their eggs among algal mats while performing a kind of dance with the male.

Distribution: From coast to coast within the United States.

Remarks: Length = 55 mm. The dark markings on the base of each wing (saddlebags) have been compared to a pair of Greek comedy masks staring at each other across the dragonfly's body (Dunkle 2000).

Similar species: None of the other saddlebag dragonflies in the wetlands have a combination of black body, black saddlebags, and white markings on the abdomen.

Fig. 2-15
Red-mantled
glider (*Tramea
onusta*)

Red-mantled glider
Tramea onusta
(FIG. 2-15)

Biology: Red-mantled gliders prefer quiet waters. Males seem to rest high in the trees as well as in the lower vegetation that is so commonly used by other dragonflies. They patrol large territories and mate with females while perched (Dunkle 1989). Mated females lay their eggs on algal mats alone or while still coupled to their mates. This species and its closest relatives are known as dancing gliders because of the dipping flight pattern they adopt. According to Dunkle, the dark red bands at the base of the hind wings provide shade on hot days when the dragonfly droops its abdomen beneath the dark patches. These have given the species an alternative common name of red saddlebags.

Distribution: From Indiana to the Pacific Ocean.

Remarks: Length = 49 mm. Red-mantled gliders are more difficult to catch than any of the other dragonflies with the exception of their relative the black saddlebags. They cruise along in a bouncing flight, tend to stay farther from shore, and are prone to speed away from the pond entirely when the net swings and misses. However, they are not common.

Similar species: The violet-masked glider (*T. carolina*) might occur in the area. It has more red in the hind wing, and the outline of the dark wing pattern is not so jagged as that of the red-mantled glider.

Fig. 2-16
Black setwing
(*Dythemis nigrescens*)

Black setwing
Dythemis nigrescens
(FIG. 2-16)

Biology: Black setwings are very slender dragonflies that frequent a broad spectrum of moving and stagnant waters. They forage in shady places, darting out from a perch to capture their prey. Males patrol extensive territories (Dunkle 2000).

Distribution: Within the United States this species would be confined to Texas were it not for a disjunct population in southwestern Arizona. The Texas distribution lies within the southwestern quadrant of the state, and according to the map of Dunkle (2000), our discovery of the black setwing in the Ottine wetlands marks a new eastern record.

Remarks: Length = 45 mm. This is one of many dragonflies that we encountered in the low-lying swamps but never saw during years of study in the nearby Lost Pines forest. There seems to be little published information regarding its habits.

Similar species: Among dragonflies with dark males only the slaty skimmer is likely to be confused with this species, but the former is much larger and does not have a slightly clubbed abdomen.

Green darner
Anax junius
(FIG. 2-17)

Biology: Adults are common in Soefje Swamp, where males patrol territories and mate with less brightly colored females. The female typically lays her

Fig. 2-17
Green darner
(*Anax junius*)

eggs in tissues of submergent vegetation, often while still coupled to the male, and she sometimes submerges herself as she does so. This is in sharp contrast to the behavior of those species that simply drop the eggs in water.

Like other dragonflies green darners are diurnally active predators that consume huge numbers of mosquitoes and other small flying insects. There are two reports of attacks on hummingbirds (Dunkle 1989), and flocks of darners have been seen flying in pursuit of insect prey (Walker 1958). Juveniles, like those of all other dragonflies, are also predatory, but they catch their prey on the bottom of the pond rather than in the skies. They are wingless and equipped with a complicated set of mouthparts known as "the mask," which flicks forward like a chameleon's tongue to catch passing animals as large as tadpoles. The structure and use of the mask has resulted in its comparison to a combination of hands, carving tools, and serving table (Needham and Westfall 1954). In July darner larvae shed their skins, which cling to the stems of aquatic plants above the waterline.

Distribution: The green darner occurs throughout the United States.

Remarks: Length = 80 mm. This is one of the largest dragonflies in the United States, though it is a bit smaller than the rarely seen swamp darner, which also occurs in the Ottine wetlands. When dozens of big green darners fly back and forth just above the waterline in the heart of a green ash swamp, they seem to provide a window to the world as it was 100 million years ago.

Similar species: Large size in combination with blue and green colors, especially conspicuous in the male, distinguishes green darners from all other species.

Fig. 2-18
Swamp darner
(*Epiaeschna heros*)

Swamp darner
Epiaeschna heros
(FIG. 2-18)

Biology: The swamp darner's habits are similar to those of the green darner, which it slightly exceeds in size. It does seem to wander farther from the ponds than its close relative and is occasionally captured in buildings (Walker 1958). We saw them flying and roosting in shaded areas along tree lines. Adults feed in swarms on winged ants and termites and on other prey as large as cicadas (Dunkle 1989). Males do not defend territories. They mate with the female while hanging from a tree. The female does not lay her eggs while attached to the male, as the green darner does, nor does she lay her eggs below the waterline. Instead, she chooses emergent vegetation or even the soil of a dry pond. In late June in conditions that included intermittent rains, we watched several females laying eggs in soggy branches on the forest floor in dry lagoons.

Distribution: From the Atlantic Ocean to a western limit in northern Texas and midwestern states to the north.

Remarks: Length = 91 mm. Unlike the green darner, this big dragonfly loses its attractive colors shortly after death.

Similar species: None.

Fig. 2-19
Banded dragonlet
(*Erythrodiplax
umbrata*)

Banded dragonlet
Erythrodiplax umbrata
(FIG. 2-19)

Biology: Adult banded dragonlets are often seen in green vegetation around small ponds. Males are more likely to be noticed than females because of their territorial behavior and because females often rest in trees. Mating occurs on the wing, but the male does not guard the female when she lays her eggs in the water.

Distribution: In the United States the distribution is very limited, occurring only in Texas and Oklahoma.

Remarks: Length = 45 mm. Young males do not have dark bands on their wings, and in this they resemble females. As males age, the bands appear and become dark in color.

Similar species: None.

Dwarf dragonlet
Erythrodiplax minuscula
(FIG. 2-20)

Biology: This tiny species preys upon damselflies. Mating takes place in the air or while the partners perch on vegetation, and males guard females while they lay their eggs in water (Dunkle 1989).

Distribution: From the Atlantic Ocean to a western limit in Texas near the Ottine wetlands.

Remarks: Length = 27 mm. The dwarf dragonlet and the eastern amberwing

Fig. 2-20
Dwarf dragonlet
(*Erythrodiplax
minuscula*)

are the smallest dragonflies we encountered. The dragonlet is likely to be overlooked, as it hides in lush green vegetation along the edge of a pond. Its secrecy undoubtedly saves it from larger predatory relatives.
Similar species: None.

Fig. 2-2
Blue dasher
(*Pachydiplax
longipennis*)

Blue dasher
Pachydiplax longipennis
(FIG. 2-21)

Biology: This is an abundant, medium-sized, fast flyer with a wide habitat tolerance. It frequents still waters with or without fish, including ponds, marshes, bays, ditches, and swamps (Dunkle 2000). Perhaps the larval stage does not do well in the more acidic waters of wetlands farther east than these, for the species is said to avoid bogs. When resting on a stem, the territorial male raises its abdomen high in the air, thus performing a headstand.

Distribution: Coast to coast within the United States except for the Great Basin region.

Remarks: Length = 45 mm. Blue dashers are common at artificial ponds and in the Soefje Swamp in early May.

Similar species: Male eastern pondhawks bear a vague resemblance to the male blue dasher, but they lack the brownish patches on the distal half of the wing.

Fig. 2-22
Common bas-
kettail
(*Epitheca
cynosura*)

Common baskettail
Epitheca cynosura
(FIG. 2-22)

Biology: This medium-sized species flies fast and erratically and includes winged termites in its diet (Dunkle 1989). It strays far from its native pond when foraging for prey. Males patrol up to about thirty feet of shoreline and mate with females in flight. The females deposit the eggs in an unusual way, for they are not laid individually but form a ball-shaped mass that unravels underwater as a gelatinous rope.

Distribution: From the Atlantic Ocean to Wyoming.

Remarks: Length = 43 mm. The common name baskettail refers to the

Fig. 2-23
Dot-winged
baskettail
(*Epitheca
petechialis*)

female's habit of carrying her eggs in a ball-shaped mass beneath her abdomen.

Similar species: The dot-winged baskettail (*E. petechialis;* length = 48 mm; Fig. 2-23), a species that is considered by some to be the same as *E. costalis,* also occurs in the wetlands.

Eastern pondhawk
Erythemis simplicicollis
(FIG. 2-24)

Biology: This widespread dragonfly is voracious even for its kind and eats grasshoppers, damselflies, deer flies, butterflies, and other dragonflies (Walker and Corbet 1975; Needham and Westfall 1954; Dunkle 1989). It watches large animals, including humans, as they move through vegetation and attacks prey as they are flushed out (Dunkle 2000). Mating usually occurs on a perch rather than in midair, and females are sometimes guarded by their mates as they lay eggs in the water. Males confront each other aerially in a kind of circular treadmill display. One male hovers in front of the other for a moment before dropping down to yield the front spot to the second male, which soon reciprocates, restoring the original alignment.

Distribution: From the Atlantic Ocean to Arizona.

Remarks: Length = 44 mm. Its attractive color has given the eastern pondhawk the alternative name greenjacket. This species and the blue dasher are especially abundant in the wetlands.

Similar species: None.

Fig. 2-24
Eastern pond-
hawk (*Ery-
themis
simplicicollis*)

Fig. 2-25
Great pond-
hawk *(Ery-
themis
vesiculosa*)

Great pondhawk
Erythemis vesiculosa
(FIG. 2-25)

Biology: Great pondhawks prefer ponds and other bodies of still water. They
are known as voracious predators even among dragonflies and attack butterflies as well as their own kind (Needham and Westfall 1954). Males
patrol sizable territories and mate with females while perched on vegetation. Females lay their eggs in the water by tapping the surface with the
tip of the abdomen.

Distribution: The great pondhawk is a primarily tropical dragonfly that ranges into the southern United States, where it occurs mostly in Texas and Oklahoma, though there are isolated records in several other states.

Remarks: Length = 59 mm. Before the largely grass green adult matures, the small dark patch on the front edge of the wing near the tip is also green, an unusual exception to the shades of brown and black found in most dragonflies. Great pondhawks are wary of people and difficult to catch.

Similar species: None.

Fig. 2-26
Eastern amberwing
(*Perithemis tenera*)

Eastern amberwing
Perithemis tenera
(FIG. 2-26)

Biology: Amberwings have the most complicated courtship of any North American dragonfly. They also have a very broad habitat tolerance, and the larval stage can develop in rivers, ponds, or ditches. While the adult is perching, it flexes its body and wings in a manner that suggests it is mimicking a wasp. The male selects an egg-laying site somewhere in the pond and leads a receptive female to the area. If the female decides to mate, the pair does so while perched on emergent vegetation. Then the female lays her eggs in the water while the male hovers over her as a guard against other males (Dunkle 1989). A curious feature is the explosion of the sinking egg mass (Walker and Corbet 1975), which might serve to disperse the individual eggs and decrease the likelihood of predation of the entire clutch by a single gulp from a predatory fish. Amberwings tend to fly closer to the water surface than high-flying dragonflies that might prey upon them (Needham and Westfall 1954).

Distribution: From the Atlantic Ocean to Arizona.

Remarks: Length = 25 mm. This is one of the smallest dragonflies in the wetlands. Males are easily recognized by their orange wings with a few darker spots on each. Females have lighter wings with more spots or bands.

Similar species: None.

Fig. 2-27
Halloween
pennant
(*Celithemis
eponina*)

Halloween pennant
Celithemis eponina
(FIG. 2-27)

Biology: In its fluttering flight and in its beautiful wing coloration this species is often compared to a butterfly. We found Halloween pennants at one large artificial pond and rarely if ever sighted it elsewhere. The female lays her eggs in the water in the morning, and she does so while still in the grasp of her mate (Dunkle 1989). Halloween pennants fly in the rain and in strong wind (Needham and Westfall 1954).

Distribution: From the Atlantic Ocean to a western extreme in New Mexico.

Remarks: Length = 42 mm. The common name is a good one, for the wings are entirely orange-yellow with the exception of the bands, which are dark brown rather than black.

Similar species: None.

Fig. 2-28
Military clubtail
(*Gomphus militaris*)

Military clubtail
Gomphus militaris
(FIG. 2-28)

Biology: Military clubtails prefer still water or waters that move slowly at best. They are also among the dragonflies that perch on the ground as well as in vegetation (Dunkle 2000). We encountered a mating pair in flight in early June.

Distribution: In the United States this dragonfly occurs from South Dakota to Mexico in a corridor consisting of the plains states and all of Texas except the extreme eastern edge.

Remarks: Length = 53 mm. Another common name is sulphur-tipped clubtail.

Similar species: Clubtails are diverse and difficult to identify. A useful character in the case of male military clubtails appears to be the presence of a rounded yellow dot on the dorsum near the tip of the abdomen.

Five-striped leaftail
Phyllogomphoides albrighti
(FIG. 2-29)

Biology: Five-striped leaftails breed in flowing water, such as the San Marcos

Fig. 2-29
Five-striped leaftail (*Phyllogomphoides albrighti*)

River and Rutledge Creek. Males are territorial and rest on the ground as well as on vegetation.

Distribution: Within the United States this species would be confined to Texas were it not for a large and curiously disjunct population in western New Mexico. It is best characterized as a western species, unlike most of the plants and animals in these wetlands.

Remarks: Length = 63 mm. Unfortunately, we did not see the closely related blue-faced ringtail (*Erpetogomphus eutainia*), though here in central Texas it is known only from a handful of counties, including Gonzales County of the Ottine wetlands (Dunkle 2000).

Similar species: The four-striped leaftail (*P. stigmatus*) has a similar range, though we did not see it. A distinguishing feature is the lack of a particular stripe on the thorax that is present on the five-striped species (see Dunkle 2000).

Flag-tailed spinyleg
Dromogomphus spoliatus
(FIG. 2-30)

Biology: Favored habitats include wetlands where muddy river or pond bottoms are available for the larval stage. Perching adult males raise the abdomen vertically when overheated, thus minimizing the exposure of the body to direct sun. This maneuver is called "obelisking" (Dunkle 2000).

Fig. 2-30
Flag-tailed
spinyleg (*Dromogomphus spoliatus*) showing hind legs where spines are located

Distribution: From Maryland to Texas.

Remarks: Length = 65 mm. This species does not seem to be especially common in the wetlands.

Similar species: The black-shouldered spinyleg (*D. spinosus;* length = 68 mm) might occur in the Ottine wetlands, though we did not see it. The black-shouldered spinyleg has a broad black band in the shoulder region that the flag-tailed species lacks.

Smoky rubyspot damselfly
Hetaerina titia
(FIG. 2-31)

Biology: Rubyspots are damselflies rather than dragonflies, though they are large enough to be mistaken for the latter. Favored habitats are the flowing waters of rivers and streams (Dunkle 1989), and we therefore found both sexes along the banks of the San Marcos River. The diet includes other rubyspots. The female is remarkable for her egg-laying behavior. She submerges herself in running water to depths approaching one meter and then lays her eggs in either living or dead vegetation. Meanwhile, her mate hovers overhead, waits for her return, and guards the site against other insects.

Distribution: From coast to coast within the United States.

Remarks: Length = 44 mm. These are among the first damselflies or dragon-

Fig. 2-31
Smoky
rubyspot
damselfly
(*Hetaerina
titia*)

flies to appear, showing up in early March when the transition from winter to spring has not officially been made.

Similar species: The American rubyspot (*H. americana*) might occur in the
wetlands. The American species never has an entirely black hindwing. We
thank Dr. John Abbott for pointing out this disctinction.

3
Butterflies and Moths

Graceful flight and attractive appearance have earned butterflies a following comparable to that enjoyed by the birds. The moths, their close relatives, are not as greatly appreciated, but the tortricid moth (*Pammene medioalbana*) is notable for being discovered and made known to science from specimens collected in Palmetto State Park (Knudson 1986). Until its discovery in the Ottine wetlands no new species of its genus had been described for sixty years.

Zebra
Heliconius charitonius
(FIG. 3-1)

Biology: We found a single specimen of this beautiful tropical butterfly flitting in the shade near a hillside seep in the Soefje peatlands. Some authors describe the zebra's flight as weak, though others describe it as agile. According to our experience they maneuver well but are not particularly fast on the wing. The specimen was netted only a few feet from a clearing filled with flowering peppervine. Several Julia butterflies in search of nectar were visiting these plants (see discussion of Julia butterfly in next section). Presumably, the closely related zebra had been nectaring at the same source. It was a hot July midafternoon with a temperature just outside the shaded tree line of about 100°F. In the Houston area southeast of the wetlands, lantana flowers are the favored nectar source (Tveten and Tveten 1996).

Males recognize the female before she leaves the chrysalis, wherein the change is made from caterpillar to adult, and they attempt to mate with her before she escapes from the pupal enclosure. When mating is com-

Fig. 3-1
Zebra (*Helico-nius charito-nius*)

pleted, the female lays her eggs singly or in small groups on the passion-vine (*Passiflora* spp.), which her caterpillar brood consumes.

Poisons obtained from passionvine leaves might explain the relative immunity of adults to predation, although some botanists speculate that the animal itself might produce protective toxins.

Distribution: The zebra is best described as a tropical butterfly that ranges north to the southern half of Texas with summer strays reaching into lati-tudes as high as the midwestern United States.

Remarks: Wingspan = 86 mm. Like many bats, zebras roost in groups, and the comparison is strengthened by their habit of hanging upside down while doing so. However, like all or nearly all butterflies, they are diurnally active, and thus the zebra roosts at night rather than during the day.

Similar species: None.

Julia
Dryas iulia
(FIG. 3-2)

Biology: Julias fly back and forth along "traplines" in search of nectar. Traplines are paths that are used repeatedly, a strategy quite different from the seemingly random fluttering that some butterflies engage in while looking for food. We found Julias patrolling large sunlit patches of pepper-vine on a blistering July afternoon when the temperature in the clearing reached 100°F. An account of the species from nearby Houston reported activity only from September through November (Tveten and Tveten

Fig. 3-2
Julia (*Dryas iulia*)

1996), and under these more urban conditions lantana flowers were the preferred source of nectar. Males patrol for females as well as food, and once mated, the female lays her eggs singly on passionvine, the only food plant acceptable to the Julia's caterpillar stage.

Distribution: Like the zebra, the Julia is a tropical butterfly that ranges into south Texas and strays farther north in summer.

Remarks: Wingspan = 96 mm. The Julia's search for moisture and minerals leads it to feed in groups at rain puddles and even at the eyes of basking turtles and South American crocodilians (Tveten and Tveten 1996).

Similar species: The Gulf fritillary (Fig. 3-3) has distinct silver patches on the lower surface of its wings.

Gulf fritillary
Agraulis vanillae
(FIG. 3-3)

Biology: Adults sip wildflower nectar. The caterpillars feed on passionflower leaves.

Distribution: From coast to coast within the United States.

Remarks: Wingspan = 70 mm. Like the giant swallowtail the Gulf fritillary has the look of the Tropics about it, as does the Julia and the closely related black and yellow zebra butterfly (*Heliconius charitonius*).

Similar species: None.

Fig. 3-3
Gulf fritillary
(*Agraulis vanillae*)

Fig. 3-4
Pipevine swallowtail (*Battus philenor*) captured by audacious jumping spider (*Phidippus audax*)

Pipevine swallowtail
Battus philenor
(FIG. 3-4)

Biology: Adult pipevine swallowtails are common in spring and to a lesser extent in summer when they flit among wildflowers in search of nectar. Males also fly in search of mates, and once mated, the females seek out host plants on which to lay their small orange eggs. Mating pairs seem unconcerned with their surroundings and can be approached and captured by hand. Caterpillars hatch from the eggs and become abundant on the sand in warning colors of red or purplish black. These colors discourage at least some potential predators from eating the caterpillar, which contains toxins acquired while consuming leaves of the eerily beautiful pipevine plant, also known as swanflower. A second defense exists in the form of the osmeterium structure, which is diagnostic for swallowtail caterpillars. This is a yellow, forked organ that appears suddenly from just behind the head when the caterpillar is disturbed. The osmeterium can startle a foe by its appearance and by the odor of a defensive chemical that stains human skin yellow if it touches the organ. We found the odor rather pleasant, which is not paradoxical because the osmeterium did not evolve as a defense against humans.

Distribution: From the Atlantic Ocean to El Paso, Texas, with isolated populations farther west between New Mexico and the Pacific Coast.

Remarks: Wingspan = 110 mm. When the caterpillar creeps across the soil, it appears to be probing its environment with two large antennae. These are actually fleshy stalks attached *behind* the head and thus are not true antennae at all. The larva's true antennae are too small to be noticed. As we saw no swanflowers, the caterpillars are presumably eating an alternative host plant, perhaps water-pepper (*Polygonum hydropiperoides*). Other names for the pipevine swallowtail are blue swallowtail, green swallowtail, and Aristolochia swallowtail.

Similar species: The pattern of conspicuous orange spots on the lower surface of the hind wing distinguishes the adult pipevine swallowtail from the similar but less abundant black swallowtail. Black swallowtails also have spots, but they are arranged in a different pattern and have an orange spot on top of each hind wing that the pipevine lacks.

Black swallowtail
Papilio polyxenes
(FIGS. 3-5–3-6)

Biology: Adults sip nectar from wildflowers, whereas the green and black caterpillar stage feeds on plants of the parsley family.

Distribution: Black swallowtails range from the Atlantic Ocean to the Rocky Mountains.

Remarks: Wingspan = 89 mm. This butterfly flies faster and higher than the pipevine swallowtail that it closely resembles. It is less common than the pipevine species and less approachable.

Similar species: See pipevine swallowtail.

Fig. 3-5
Black swallow-
tail (*Papilio
polyxenes*)

Fig. 3-6
Black swallow-
tail caterpillar
reacting to
disturbance by
displaying the
fork-shaped
osmeterium

Giant swallowtail
Papilio cresphontes
(FIG. 3-7)

Biology: Adults are very large brown and yellow butterflies that fly high, are
difficult to catch with a net, and are seldom seen, though we saw them in
the Ottine swamps, especially in the heat of summer, more often than
elsewhere in central Texas. They do have the look of the Tropics about
them. Brown and buff caterpillars resemble bird droppings and are known
as orange dogs or orange puppies, but not because of any colors they dis-

Fig. 3-7
Giant swallow-
tail (*Papilio cresphontes*)

play. They received these common names because they feed on members of the citrus tree family, including the hop-tree, which must be an important host located along the interface between the wetlands and the adjacent uplands. The odor of the orange dog's osmeterium is more offensive to humans than that of the pipevine swallowtail. Yet it does not save the larva from all enemies. On one occasion in the Lost Pines forest we returned to a hop-tree where several orange dogs were seen feeding on leaves the previous week. All were gone and presumably eaten.

Distribution: From the Atlantic Ocean to central Texas not far from the Ottine swamps, with an isolated population in California.

Remarks: Wingspan = 140 mm. This is the largest butterfly in the region. It is peculiar in its habit of flitting rapidly through an area instead of lingering as the other swallowtails often do and for its tendency to fly too high for capture by a net. The caterpillar is also exceptional for its protective bird-dropping coloration that is so unlike that of the other swallowtails when they approach maturity.

Similar species: None.

Eastern tiger swallowtail
Papilio glaucus
(FIG. 3-8)

Biology: We first encountered this beautiful species in mid-March as it nectared from the brightly colored, newly opened tubular flowers of red buckeye (*Aesculus pavia*). Tigers also favor buttonbush (*Cephalanthus occi-*

Fig. 3-8
Eastern tiger swallowtail (*Papilio glaucus*)

dentalis), a small tree or shrub that flowers later in the year. In and about the summer-dried lagoons of Palmetto State Park, tigers and pipevine swallowtails can be seen competing for buttonbush nectar.

   The caterpillar stage feeds upon the foliage of many plants, but ash trees, which are abundant in the Ottine swamps, are one of their favorites. Other host plants include willows, elms, grapes, and oaks (Tietz 1972).

Distribution: There is disagreement regarding the extent of the eastern tiger swallowtail's range. Perhaps some disagreement is due to confusion with the western tiger swallowtail, which we did not see, or at least identify, in the Ottine swamps. In the United States eastern tiger swallowtails are said to occur from the Atlantic Ocean to the Rocky Mountains. In the northern states they occur farther west. Though it is known to frequent forests and had been reported by others, we never saw this swallowtail during years of fieldwork in the nearby Lost Pines forest.

Remarks: Wingspan = 140 mm. The eastern tiger swallowtail is the most famous butterfly in the United States. In 1587 it became the first American species to be illustrated by a newly arrived English colonist.

Similar species: No other tiger-striped swallowtail is likely to be seen in this region. Some females are dark and can be confused at a distance with pipevine swallowtails. Close examination of a dark tiger will reveal its stripes.

Fig. 3-9
Question mark
(*Polygonia
interroga-
tionis*)

Fig. 3-10
Question mark
caterpillar

Question mark
Polygonia interrogationis
(FIGS. 3-9–3-10)

Biology: Adults eschew flower nectar in favor of tree sap, decaying fruit, and dung. Caterpillars feed on elm and hackberry leaves, both trees being abundant in the wetlands.

Distribution: From the Atlantic Ocean to the Rocky Mountains.

Remarks: Wingspan = 71 mm. This is one of the most common butterflies of spring, but it becomes scarce as summer's heat intensifies. The lower surface of each wing resembles a dead leaf. The lower surface of the hind wing in particular also bears a small silvery design that has been compared to a question mark.

Similar species: The comma (*P. comma*) is a smaller butterfly that is similar in appearance but does not range as far south as the wetlands.

Fig. 3-11
Monarch
(*Danaus plex-ippus*)

Fig. 3-12
Queen
(*Danaus gilip-pus*)

Monarch
Danaus plexippus
(FIG. 3-11)

Biology: Adults sip nectar from wildflowers, and some individuals take part in an annual migration between Mexico and southern Canada. Many predators avoid monarchs because the attractively striped caterpillar stage feeds on toxic milkweed plants, and as a result, the poisons accumulate in its body.

Distribution: From coast to coast within the United States.

Remarks: Wingspan = 100 mm. The monarch butterfly is the official state insect of Texas, having defeated the red harvester ant for the honor.

Similar species: Monarchs have orange wings with thick black veins, whereas the closely related queen has brown wings with less prominent veins. The viceroy (*Limenitis archippus*) resembles the monarch even more closely, but we never saw a viceroy.

Queen
Danaus gilippus
(FIG. 3-12)

Biology: The queen's habits are similar to those of the monarch.

Distribution: From the Atlantic Ocean to a western limit near the Ottine wetlands.

Remarks: Wingspan = 84 mm. Neither this species nor the monarch is abundant. The queen is the more common of the two.

Similar species: See monarch.

Hackberry butterfly
Asterocampa celtis
(FIGS. 3-13–3-14)

Biology: Adults sip sap from trees, nectar from wildflowers, and liquid from rotting fruit. The caterpillars feed gregariously on leaves of the abundant hackberry tree.

Distribution: Spottily, from coast to coast across the United States.

Remarks: Wingspan = 66 mm. A close relative, *A. clyton,* is also a hackberry butterfly, though it is better known as the tawny emperor. Both species sometimes perch on hats or other clothing.

Similar species: Tawny emperors do not have dark eyespots on the front wing, but hackberry butterflies do.

Fig. 3-13
Hackberry
butterfly
(*Asterocampa
celtis*)

Fig. 3-14
Hackberry
butterflies
feeding on
tree sap near
cattail marsh

Fig. 3-15
Variegated frit-
illary (*Euptoi-
eta claudia*);
mating pair

Variegated fritillary
Euptoieta claudia
(FIG. 3-15)

Biology: Adults favor open areas with low vegetation, such as the small, dry clearing near a cattail marsh where we encountered a mating pair. These butterflies nectar at a variety of flowers, including milkweeds and their relatives. The caterpillar hides during the day and feeds at night on the foliage of violets, passionflowers, beggar ticks, and purslane.

Distribution: From coast to coast across the southern United States.

Remarks: Wingspan = 63 mm. *Euptoieta* means "easily frightened." This is true of the variegated fritillary, especially when the skittish mating pair is compared to that of the pipevine swallowtail, which allows the observer to take the couple in hand.

Similar species: None.

Fig. 3-16
Buckeye
(*Junonia coenia*)

Buckeye
Junonia coenia
(FIG. 3-16)

Biology: Adult buckeyes visit wildflowers for nectar, and the caterpillars feed on a wide variety of plants. Males chase other insects that enter their territories.

Distribution: From coast to coast within the United States.

Remarks: Wingspan = 58 mm. Prominent eyespots on the upper surface of the wings may serve to startle predators and thus give the adult a head start toward escape.

Similar species: None.

Red admiral
Vanessa atalanta
(FIG. 3-17)

Biology: Adults sip nectar from thistles that flower in uplands adjoining the wetlands and presumably from other flowers as well. The caterpillar feeds on a wide variety of plants.

Distribution: From coast to coast within the United States.

Remarks: Wingspan = 58 mm. Red admirals have a reputation for perching on humans. This happened to one of us on a single occasion and is indeed unusual among butterflies. At all other times they avoided contact by flying away and were difficult to approach.

Similar species: None.

Fig. 3-17
Red admiral
(*Vanessa atalanta;* left);
Hunter's butterfly
(*Vanessa virginiensis;* right)

Hunter's butterfly
Vanessa virginiensis
(FIG. 3-17)

Biology: The habits of Hunter's butterfly are similar to those of the red admiral.

Distribution: From coast to coast within the United States.

Remarks: Wingspan = 58 mm. This species, the painted lady, and the red admiral are all known as thistle butterflies because those plants figure so prominently in their diets.

Similar species: Hunter's butterfly has two large eyespots on the lower surface of the hind wing, whereas the painted lady (*V. cardui*) has four small eyespots in the same area.

Painted lady
Vanessa cardui

Biology: The painted lady's habits are similar to those of the closely related Hunter's butterfly and red admiral. Occasionally they come to black lights after dark, which is unusual behavior for a butterfly.

Distribution: From coast to coast within the United States.

Remarks: Wingspan = 58 mm. The painted lady is the world's most widely distributed butterfly.

Similar species: See Hunter's butterfly.

Fig. 3-18
Common
snout butterfly
(*Libytheana
bachmannii*)

Common snout butterfly
Libytheana bachmannii
(FIG. 3-18)

Biology: We saw the small adults flitting about in a field of wild onion flowers in late April. The caterpillar stage feeds on nothing but hackberries
(*Celtis laevigata*), trees that are abundant in the Ottine swamps, and the
species prefers river bottoms to upland habitats. Like the question mark
butterfly this one resembles a dead leaf when it perches with its wings
held vertically. Population explosions sometimes result in migrations, and
in 1966 millions of snout butterflies obscured the sun in Tucson, Arizona,
causing streetlights to turn on at midday (Opler and Krizek 1984).
Distribution: The common snout butterfly occurs from the Atlantic Ocean
to Arizona.
Remarks: Wingspan = 50 mm. This is one of the few snout butterfly species
occurring in the United States. What appears to be a long snout at the
tip of the butterfly's head is actually a set of mouthparts.
Similar species: None.

Texas crescent
Anthanassa texana
(FIG. 3-19)

Biology: Male Texas crescents are territorial and dart out from their perches
to challenge other males. Prior to mating an aerial courtship dance is per-
formed in which the male flies loops around the female (Tveten and

Fig. 3-19
Texas crescent
(*Anthanassa texana*)

Tveten 1996). The caterpillar feeds on ruellia plants, which we did notice in our travels. In summer, adults of Texas populations wander as far north as Canada.

Distribution: Despite its common and scientific names the Texas crescent occurs from coast to coast within the southern United States. The names indicate history rather than distribution, for the species was discovered in nearby New Braunfels during the Civil War (Tveten and Tveten 1996).

Remarks: Wingspan = 44 mm. Texas crescents are reasonably common. We saw adults flying in autumn.

Similar species: A few crescent butterflies and even a few checkerspots might be confused with the Texas crescent. The concavity or notch in the outer margin of the forewing is a useful aid in distinguishing this small dark butterfly from similar species.

Phaon crescent
Phyciodes phaon
(FIG. 3-20)

Biology: This butterfly's caterpillar feeds upon frog-fruit (*Phyla lanceolata*), and the adult nectars at flowers, including those of the host plant (Tveten and Tveten 1996).

Distribution: From coast to coast within the United States.

Remarks: Wingspan = 32 mm. The caterpillar stage is apparently seldom seen, and we did not notice any caterpillars.

Similar species: None.

Fig. 3-20
Phaon crescent (*Phyciodes phaon*)

Fig. 3-21
Silvery checkerspot (*Chlosyne nycteis*)

Silvery checkerspot
Chlosyne nycteis
(FIG. 3-21)

Biology: Adults sip nectar at flowers, and caterpillars feed on a variety of plants in the daisy family.

Distribution: From the Atlantic Ocean to Arizona.

Remarks: Wingspan = 44 mm. Checkerspot caterpillars feed gregariously during the early part of their development, but we never noticed them in the act.

Similar species: These butterflies and their relatives are variable in coloration and as a result are difficult to distinguish from one another.

Fig. 3-22
Falcate orangetip butterfly (*Paramidea midea*)

Falcate orangetip butterfly
Paramidea midea
(FIG. 3-22)

Biology: The orangetip is a low-flying species known to prefer damp environments, such as the floodplain of the San Marcos River. The caterpillar feeds upon the flowers of plants belonging to the mustard family but is seldom seen eating their leaves. Bitter-cress (*Cardamine bulbosa*) is a favorite host in east Texas, but this plant might not occur in the swamps. Adult orangetips are light-loving, and if a cloud obscures the sun, they are likely to hang beneath a leaf until its return (Tveten and Tveten 1996).

Remarks: Wingspan = 44 mm. This is one of the first butterflies to appear in the new year. It may be seen in winter as early as February, though it flies as an adult and feeds on mustards as a caterpillar for only a short period. Most of the year is spent in the chrysalis stage.

Distribution: From the Atlantic Ocean to Texas.

Similar species: None. The sickle-shaped (falcate) wing tips combined with the patches of mossy green on the lower surface of the wings distinguish this species from cabbage butterflies and other largely white species. Males have orange wing tips.

Fig. 3-23
Luna moth
(*Actias luna*)

Luna moth
Actias luna
(FIG. 3-23)

Biology: The spectacular luna moth is a huge, pale green moth and, like the swallowtail butterflies, sports graceful tails on its hind wings. As most moths do, these fly at night rather than during the day, especially in the midnight hour when females release sex pheromones that attract the males. Our first specimen was a male that flew to a streetlight just outside the sandstone refectory of Palmetto State Park. It arrived at eleven o'clock, in keeping with what is known about its habits, and made much noise flying between the ground and the lights overhead. Thus, we can testify to the accuracy of the claim that "males are determined fliers" (Collins and Weast 1961). On March 15 several individuals of both sexes arrived at our black lights.

   In the southern United States the luna moth has at least three broods per year, and it has been reported from Louisiana in every month (Tuskes, Tuttle, and Collins 1996). The first spring brood can be distinguished from summer generations by the purplish trim on the adult's wings. Big, green luna caterpillars eat leaves from a variety of trees and shrubs,

Fig. 3-24
Polyphemus
moth (*Antheraea polyphemus*)

including sumac, persimmon, oak, willow, and eastern cottonwood. Upon
completion of their development within a cocoon of leaves and silk, they
use a claw-shaped appendage at the base of each forewing to make their
escape. The sawing, cutting sounds can be heard from twenty feet away
(Priddle 1967; see photos therein).

Distribution: In the United States the luna moth occurs from the Atlantic
Ocean to a western limit near San Antonio, Texas.

Remarks: Wingspan = 114 mm. This nocturnal moth is arguably more
attractive than the prettiest butterflies that fly in full sun. The literature
gives an impression that the species is common in the eastern half of
Texas, but our experience does not support that generalization. Despite
decades of experience with black lights and porch lights our first
encounter with a wild adult was the specimen from the Ottine swamp pictured here. It is acknowledged that lunas become less common across their
range from east to west and that they are more rural than some of the
other large native silk moths that occasionally fly to porch lights in the
city.

Similar species: None.

Polyphemus moth
Antheraea polyphemus
(FIG. 3-24)

Biology: Polyphemus caterpillars are said to favor leaves of water oak, a tree
that prefers moist areas but seems to be rare if it occurs at all in these
wetlands. Other oaks, hickory, elm, and plum are also accepted. We saw
more adults than we did of any other large silk moth, both sexes winging

Fig. 3-25
Robin moth
(*Hyalophora cecropia*)

their way to our black lights above the San Marcos River, especially in mid-March.

Distribution: From the Atlantic Ocean to the Rocky Mountains.

Remarks: Wingspan = 140 mm. It is notable that while blacklighting in Lost Maples State Park in west-central Texas on a single night, we captured this species and the robin moth, but neither appeared at our traps in the Lost Pines even though we set them out over the course of several years. It is possible that populations are being reduced by a fly imported into the United States to control gypsy moths. The fly, *Compsilura concinnata,* was introduced in 1906 for this purpose only, but there is evidence that it is killing native silk moths as well (Jensen 2000).

Similar species: None.

Robin moth
Hyalophora cecropia
(FIG. 3-25)

Biology: This is the largest moth or butterfly in the United States. Its caterpillar feeds upon a wide variety of trees and bushes, including the black willow, where we found the cocoon of a male specimen in late summer. In Texas the moth might leave its overwintering/maturation container as early as March of the following year if a rodent does not find it, eat the contents, and use the empty shell as a storage bin for seeds (Collins and Weast 1961). Woodpeckers eat them too (Tuskes, Tuttle, and Collins 1996). Other host plants that grow in the swamp and its immediate surroundings are boxelder, buttonbush, ash, elderberry, pecan, and oak. Male moths seek out females by following an airborne chemical pheromone that the latter secretes. After mating, the female lays eggs on the leaves of a host that is suitable to the caterpillar stage. We attracted two females to our black lights on March 15, as well as several individuals of both sexes of the luna and polyphemus moths, and conclude that this is the best time of year to see them.

Distribution: From the Atlantic Ocean to Alberta with a Texas western limit near Kerrville.

Remarks: Wingspan = 160 mm. Our single capture of an adult at a black light in Lost Maples State Park near Kerrville appears to be a western record for Texas. Another, more familiar common name for the robin moth is cecropia moth.

Similar species: None. The enormous size and deep red colors distinguish this species from related silk moths.

Black witch
Ascalapha odorata
(FIG. 3-26)

Biology: We spied the big male moth pictured here as it flew in the deep shade of trees growing along the bank of Rutledge Creek. It landed on a trunk just within reach of the net, and though it escaped the first swing, it made a wrong turn soon after and was bagged with a second attempt.

Adult black witches have a pair of ears complete with eardrums on the thorax just in front of the abdomen. They probably use these while flying

Fig. 3-26
Black witch
(*Ascalapha*
odorata)

to detect the sonar signals of predatory bats, thus giving them a head start and a chance to evade enemies that they must sometimes rival in size. Introduced locust trees and the introduced fig occur near the Ottine swamps, but we did not notice these or any of the other less likely food plants that might be consumed by the caterpillar stage. Perhaps the moth flew in from the south or west.

Distribution: The black witch is a tropical species that ranges from South America to northern limits in southern Texas and Florida.

Remarks: Wingspan = 152 mm. This is a huge, dark, otherworldly-looking moth. Large numbers flew ashore in Matagorda Bay on July 15, 2003, in the eye of Hurricane Claudette as it made landfall on the Texas Gulf Coast. The moths drank beer from the cupped hands of observers who chose to ride out the storm and its 110-mile-per-hour winds (Freeman 2003).

Similar species: Large size and dark brown color easily distinguish the black witch from all other species in the swamp.

Fig. 3-27
Sad underwing
(*Catocala maestosa*)
camouflaged on bark

Sad underwing
Catocala maestosa
(FIG. 3-27)

Biology: Big adult moths are abundant in late spring and less so in summer but are seldom noticed unless disturbed during daylight hours because they rest on the trunks of ash trees, where they blend in magnificently with the color and furrows of the bark. One prominent black bar on each forewing contributes to the effect, for it seems to line up with the bark's deep, shadowed grooves no matter where the moth happens to perch. The sad underwing's caterpillar stage feeds upon the foliage of the pecan trees that are so abundant near the swamp, though the adults seem to favor resting spots on green ash.

Distribution: From the Atlantic Ocean to a western limit in Texas.

Remarks: Wingspan = 98 mm. According to one source the sad underwing is generally uncommon, though it can be locally common in the south (Sargent 1976). This is certainly supported by our experience in the Ottine swamp.

Similar species: None that we have seen.

Mesquite cutworm moth
Melipotis indomita
(FIG. 3-28)

Biology: This small, drab, grayish, and presumably nocturnal moth arrived at our Malaise trap during daylight hours in the first week of March, the earliest appearance of the flying adult stage that we are aware of. Its caterpillar is highly significant as the most important defoliating enemy of mesquite (*Prosopis glandulosa*) in central Texas (Cuda, DeLoach, and Robbins 1990; DeLoach 1994). Young larvae feed only on immature leaves and do not hide on the ground during the day as much as older larvae do. As the caterpillars age and grow, they eat older leaves, and they leave their nocturnal feeding sites as day approaches to hide in cover beneath the tree. These larger individuals are presumably protecting themselves from diurnal predators such as birds and wasps.

Distribution: From the Atlantic Ocean to at least as far west as Arizona.

Remarks: Wingspan = 50 mm. This moth has been considered for use as a biological control agent of mesquite, but the species would probably prove unequal to the task despite the massive defoliation that the rangeland trees endure (DeLoach 1994). In Palmetto State Park mesquite groves can be reached within a few minutes' walk from ash swamp and cattail marsh.

Fig. 3-28
Mesquite cut-worm moth (*Melipotis indomita*)

Similar species: None that we have seen. The wing pattern as illustrated here is unique.

Yellow-striped cutworm
Spodoptera ornithogalli
(FIG. 3-29)

Biology: This attractive caterpillar, also known as the cotton cutworm, is an economically important pest of many unrelated crops. The adult moth is brown, gray, yellow, and white, though we did not notice any.
Distribution: At least sporadically throughout the United States.
Remarks: Length = 44 mm. The specimen shown here was photographed while, unseen by either author, the juvenile water moccasin of Fig. 1-33 was poised to strike the photographer's leg. As it turned out, only the caterpillar followed through on its threat, squirting a green jet of fluid that found its mark when the camera approached too closely.
Similar species: None.

Nessus sphinx moth
Amphion floridensis
(FIG. 3-30)

Biology: In flight the nessus sphinx resembles a large wasp. Its caterpillar stage feeds on grape leaves and Virginia creeper, both vines growing abun-

Fig. 3-29
Yellow-striped cutworm (*Spodoptera ornithogalli*)

Fig. 3-30
Nessus sphinx moth (*Amphion floridensis*)

dantly in the swamps. Adults sip nectar from a variety of flowers, including species of phlox, perhaps including the brilliant red Drummond's phlox (*Phlox drummondii*).

Distribution: The nessus sphinx occurs from the Atlantic Ocean to the Rocky Mountains.

Remarks: Wingspan = 55 mm. We saw a single individual. It was flying in late afternoon just aboveground at a tree line along the southern edge of the dry center of Soefje Swamp. Until it was netted, we had mistaken the nessus sphinx for a spider wasp.

Similar species: None.

Fig. 3-31
Carolina
sphinx moth
(*Manduca
sexta*)

Carolina sphinx moth
Manduca sexta
(FIG. 3-31)

Biology: The big, beautiful Carolina sphinx moth came to our black light above the San Marcos River. The massive green caterpillar, known as the tobacco hornworm, bears a red-tipped horn near the end of its body and feeds upon potato, tomato, and tobacco plants. In the Ottine swamps it might eat leaves of the Texas nightshade (*Solanum triquetrum*). Pupation occurs underground without a cocoon.

Distribution: From coast to coast within the United States.

Remarks: Wingspan = 120 mm. The caterpillar sometimes becomes a pest in gardens. Sphinx moths are also known as hawk moths.

Similar species: The related tomato hornworm or five-spotted hawk moth (*M. quinquemaculata*) may be distinguished in the adult stage by the bands of the hind wing. The center two bands are not fused as they are in the Carolina sphinx (Covell 1984).

Waved sphinx moth
Ceratomia undulosa
(FIG. 3-32)

Biology: Adults appeared in numbers at our black lights during the first week of April. The caterpillar feeds on the leaves of ash trees, oaks, hawthorn, and the trumpet creeper vine. We saw a single caterpillar on ash in Palmetto State Park.

Fig. 3-32
Waved sphinx
moth (*Cerato-
mia undulosa*)

Distribution: From the Atlantic Ocean to a western limit in western Texas or perhaps eastern New Mexico.

Remarks: Wingspan = 110 mm. This moth has an unusually thick pile of scales on its thorax. These wear off, leaving an unsightly bare spot when the specimen is placed haphazardly in standard Lepidoptera envelopes.

Similar species: None.

Great leopard moth
Ecpantheria scribonia
(FIG. 3-33)

Biology: These large, attractive moths flew to our black lights above the banks of the San Marcos River in spring. Their white wings bear leopard-like spots that give the animal its common name. The hairy black caterpillar exceeds 50 mm in length and bears red coloration between its segments that becomes especially apparent when the larva rolls into a ball,

Fig. 3-33
Great leopard
moth (*Ecpan-
theria scribo-
nia*)

perhaps warning potential predators of unpalatability. Its food in the wetlands includes willow, boxelder, and the low-growing violets so abundant
along Rutledge Creek. Pupation occurs beneath logs or bark, and on one
occasion the claim was made that these larvae seek sites where ants are
abundant (Knetzger 1908). Perhaps the ants discourage potential predators of the moth during its vulnerable period of transformation. One successful predator of the adult and presumably the caterpillar is the fiery
searcher (Fig. 4-37).

Distribution: From the Atlantic Ocean to Texas.

Remarks: Wingspan = 91 mm. Fast-moving caterpillars are sometimes seen
crossing roads in Palmetto State Park. Dead adults often turn up in urban
parking lots because of their attraction to lights.

Similar species: None.

Salt marsh moth
Estigmene acrea
(FIG. 3-34)

Biology: This attractive salt-and-pepper moth flies at night and is named for
one of its larval haunts. We found several adults resting at light fixtures
and at our black lights after dark. Unlike the luna they were sometimes
inactive and did not fly even when tapped with the handle of an insect
net. They simply fell to the ground like stones. Perhaps the low nighttime
temperature of roughly 50°F accounted for their lethargy. Salt marsh
caterpillars eat an enormous variety of herbaceous plants, including a

Fig. 3-34
Salt marsh moth (*Estigmene acrea*)

Fig. 3-35
Forest tent caterpillars (*Malacasoma disstria*) on ash bark

good portion of those grown as garden vegetables. These include onion, lettuce, peas, and beans.

Distribution: From coast to coast within the United States.

Remarks: Wingspan = 68 mm. Males are readily distinguished from females by their yellow-orange hind wings. Females have white hind wings instead.

Similar species: None.

Forest tent caterpillar
Malacasoma disstria
(FIG. 3-35)

Biology: In the wilds of the Ottine wetlands the forest tent caterpillar feeds upon the leaves of green ash, boxelder, oaks, willows, and elm. Green ash

Fig. 3-36
American dagger moth caterpillar (*Acronicta americana*)

Fig. 3-37
Cattail moth (*Dicymolomia julianalis*) with pupal skin (center)

is the flagship tree of the swampland and happens to be a particular favorite of the immature insect. The common name is misleading because these caterpillars do not make and live within tents as their close relatives do. Adults are drab reddish yellow moths that draw no special attention to themselves. Females lay their eggs in bunches on tree branches, including trees that are not known to be food for their young. When the caterpillar is ready to transform, it does so in a cocoon attached to a leaf.

Distribution: From coast to coast within the United States.

Remarks: Wingspan = 37 mm. The caterpillar's attractive blue color is not conferred by a pigment but by tiny structures that reflect blue light especially well because of their size, shape, and underlying layer of black pigment (Fitzgerald 1995).

Similar species: None.

American dagger moth
Acronicta americana
(FIG. 3-36)

Biology: The very large, hairy, yellow and black caterpillar is a common sight in spring. In the Ottine wetlands it feeds upon the leaves of green ash, elms, oaks, and willow. The adult is a drab gray and brown moth that we did not identify at our lights, though it was probably a frequent visitor.
Distribution: From the Atlantic Ocean to Utah.
Remarks: Wingspan = 65 mm. We think this species is parasitized by the ichneumon wasp, *Therion circumflexum* (Fig. 8-10), but we could find no suitable means of pupa/cocoon identification in the literature.
Similar species: Other dagger moths probably occur here. The most reliable identification is made during the adult stage, as surprisingly few caterpillars have been illustrated.

Cattail moth
Dicymolomia julianalis
(FIG. 3-37)

Biology: The tiny caterpillar feeds within the flower heads and stems of cattails, such as the giant cattail (*Typha domingensis*), where we found the golden cocoon hidden inside the hollow center of a dried spike remaining from a previous year's flowering period. An adult eclosed from the cocoon in culture at the end of March. Other larval foods include cactus, milkvetch, thistle, dead cotton bolls, and even the eggs of the bagworm moth (Covell 1984).
Distribution: From the Atlantic Ocean to a western limit somewhere in Texas.
Remarks: Wingspan = 20 mm. Next to the golden cocoon within the cattail stem were several small dark pupae from which tiny, yet-unidentified wasps developed to adulthood in culture. Perhaps these were parasites of another cattail moth living on the same plant.
Similar species: There are many small snout moth species in central Texas, and it is best to consult a work such as that of Covell (1984) and then to follow up with keys and the primary literature (journals).

Fig. 3-38
Carpenter moth (*Prionoxystus robiniae*)

Carpenter moth
Prionoxystus robiniae
(FIG. 3-38)

Biology: Adults fly to black lights as many other moths do, but the caterpillar is more remarkable. It does not chew leaves. It bores through wood, as a beetle grub does, after hatching from an egg laid on a tree trunk by the female, which is attracted to preexisting wounds on the trunk. Pupation also occurs inside the tree. Suitable hosts in the wetlands include ash, elm, and oaks.

Distribution: From coast to coast within the United States.

Remarks: Wingspan = 85 mm. The specimen pictured here is a female. The male's hind wings are a more colorful yellow and black.

Similar species: None.

Evergreen bagworm moth
Thyridopteryx ephemeraeformis
(FIG. 3-39)

Biology: Caterpillars spin spindle-shaped silken bags and attach bits of leaves and twigs to the protective case as they crawl about feeding on foliage. They pupate inside the bag, and the wingless female never leaves it, laying her eggs within. One of the favorite food plants is eastern red juniper (*Juniperus virginiana*), which is uncommon in the wetlands. We found caterpillars feeding on roundleaf greenbrier (*Smilax rotundifolia*) and on elderberry (*Sambuca canadensis*).

Fig. 3-39
Evergreen bag-
worm moth
caterpillar
(*Thyri-
dopteryx
ephemerae-
formis*)
exposed within
its silken bag

Distribution: From the Atlantic Ocean to Texas.

Remarks: Wingspan = 36 mm. The winged male is often pictured with transparent wings, but this transparency is due to a loss of scales during emergence from the bag where pupation takes place.

Similar species: None.

4

Beetles

Beetles are often considered to be the most successful of all animals because there are so many species (Evans and Bellamy 1996). Part of the explanation for their success might lie in the heavy coat of protective armor that most of them wear. It is probably no coincidence that the first species new to science that we encountered in central Texas was a beetle.

Predaceous diving beetle
Cybister fimbriolatus
(FIG. 4-1)

Biology: The beetle pictured here flew to our black light after dark. Both the adult and the 50 mm, gilled, aquatic larva are predators of other insects and vertebrates such as frogs and fish, occasionally becoming pests in hatcheries (Ideker 1979; Edwards 1949). Under some conditions adults scavenge dead animals exclusively (Johnson and Jakinovich 1970). The grubs, known as water tigers, inject a digestive fluid into their prey through sicklelike jaws and suck out juices, whereas adults chew their food. Mating is associated with the flight period, and after completing their development in water, the grubs crawl ashore and pupate in the soil (Johnson 1972). Males are able to produce scratching sounds by using structures associated with their hind legs.

There is one report of symbiosis resembling the behavior of cleaner-fish that remove parasites from the skin of other fish. In this case the beetle repeatedly chewed parasitic fungi from the body of an infected salamander without harming the amphibian, while the latter stood motionless (Folkerts 1967).

Distribution: At least sporadically from coast to coast but sparse in the far west.

Fig. 4-1
Predaceous diving
beetle (*Cybister
fimbriolatus*)

Remarks: Length = 33 mm. This is one of the largest predatory water beetles in the wetlands. It is well adapted for aquatic life and has been
described as among the most highly evolved species of its kind (Blatchley
1910). A new species of parasitic worm was discovered living inside the
grub stage (Poinar and Petersen 1978).

Similar species: Several large relatives, exceeding 25 mm in length and
resembling this one, are likely to occur in the wetlands. Consult formal
keys for identification.

Giant water scavenger beetle
Hydrophilus triangularis
(FIG. 4-2)

Biology: The big, shiny, convex adults are scavengers in ponds and creeks.
The gilled, cannibalistic grubs are predators of animals as substantial as
small fishes. Once the prey is captured, the beetles tear off and swallow
"large chunks" (Edwards 1949). Adult females use structures resembling
the spinnerets of spiders to make silken pouches for their eggs. The tiny
larvae inside occasionally eat one another before they have a chance to
escape from the waterproof case.

Distribution: From coast to coast within the United States.

Remarks: Length = 37 mm. This is the largest aquatic beetle in the United
States. It dwarfs the surface-dwelling whirligig but is less likely to be seen
because it remains underwater except for occasional visits to the surface
for air. Unlike the whirligig it did fly to our lights at night. Adults bear a
long, sharp spine on the lower surface that can passively spear a careless
handler.

Fig. 4-2
Giant water scavenger beetle (*Hydrophilus triangularis*)

Fig. 4-3
Oval water scavenger beetle (*Dibolocelus ovalis;* left) and the sister water scavenger beetle (*Hydrochara soror;* right)

Fig. 4-4
Greater whirligig beetle (*Dineutus ciliatus*) in Rutledge Creek

Similar species: The sister water scavenger beetle (*Hydrochara soror;* Fig. 4-3) is less than 25 mm in length, but a second relative, the oval water scavenger beetle (*Dibolocelus ovalis;* Fig. 4-3), is nearly as large as the giant species (length = 33 mm) though not so narrow in shape. It occurs at least sporadically from the Atlantic Ocean to a western limit in Texas.

Greater whirligig beetle
Dineutus ciliatus
(FIG. 4-4)

Biology: Adult whirligigs are unique or nearly so among beetles in their ability to support themselves on the film of the water's surface. They skate about scavenging on plants and animals, dive with a captured bubble of air, and fly to lights, although they never flew to ours. Each compound eye is divided into two parts so that one part looks above the water while the other looks below its surface. Thus, it appears that the beetles have four eyes rather than two. When disturbed, some species secrete a defensive chemical that smells like apples. The gilled grub stage is predatory on other aquatic insects. Greater whirligig beetles prefer creeks, streams, rivers, and ditches where running water is present, though many related species inhabit quiet, still waters such as those of the lagoons in Palmetto State Park. We found large numbers of the greater whirligig on Rutledge Creek, which flows east from the private wetlands of Ottine into the San Marcos River. Elsewhere in the United States flowing waters were also identified as the niche of this species. At lengths of about 15 mm it is the

Fig. 4-5
Rainbow dung scarab (*Phanaeus difformis*)

largest well-known whirligig in the United States (Hatch 1925).

Distribution: From the Atlantic Ocean to a western limit in Texas, perhaps near the Ottine swamps (Wood 1962).

Remarks: Length = 15 mm. The agility of these whirligigs and their favored habitat of flowing water makes their collection more challenging than the collection of pond-dwelling species. Other common names for whirligigs include waltzing beetles, scuttlebugs, and apple bugs.

Similar species: The large size and the faint bronze stripe on each wing cover separate this species from all others in the area.

Rainbow dung scarab
Phanaeus difformis
(FIG. 4-5)

Biology: The rainbow dung scarab is a mostly upland species, though it does fly down into wetland border regions where cattle graze and deposit their dung. In the Ottine area adults and grubs feed mostly upon cattle pats and, of course, the solid waste of native animals that were here long before cattle were introduced. This species is said to prefer the dung of swine and humans (Blume and Aga 1976), perhaps because omnivores have a broader diet than cattle.

Both sexes bear a horn or, at least, a tubercle on the head, though the horn of the male is much larger; and even among males the variation is great enough to warrant the recognition of two size classes. One group is the class of "majors" (long horns), and the other is the class of "minors" (short horns). The beetles use the weapon to fight other males, tipping their opponents over when the opportunity presents itself.

Though it is rare among insects, both sexes cooperate to provide dung for a burrow typically begun by a pair of beetles beneath or near the source. When males lag behind during the provisioning phase, females scratch and drum on their partners' bodies until the male returns to work (Rasmussen 1994). After a suitable amount of manure has been pushed into the burrow, the female lays a single egg, and the grub that hatches feeds upon stored dung. It is likely that the species treated here makes these brood balls in a manner similar or identical to that of a species in which the phenomenon has actually been studied. If so, the rainbow scarab's brood ball consists of an outer rind of soil, an inner shell of dung, a second layer of soil making up a thin shell, and within this innermost core an air chamber in which the egg is deposited (Halffter, Halffter, and Lopez 1974). Thus, the anatomy of the structure is more complicated than one might imagine and even suggests a comparison to the inner structure of the earth itself. This is one native species that has surely increased in numbers and benefited from human intervention, for there is probably more cow dung in the fields bordering the swamp than rainbow scarabs can use.

Distribution: This species occurs within the central plains states from Kansas to New Mexico and southern Texas and undoubtedly into northern Mexico. According to some reports, rainbow scarabs are distributed at least sporadically as far east as Florida.

Remarks: Length = 22 mm. In summer these beetles are vastly outnumbered at cowpats by the smaller, entirely black tumblebug (*Canthon imitator;* see next section). This is a large competitor that rolls dung away from the pat but does not bury it directly underneath, thus earning the common name of tumblebug.

Similar species: Close relatives of the rainbow scarab are likely to occur in the area, and they may need to be distinguished by looking at the males (Robinson 1948). Their metallic green color is also similar to that of the bumelia borer, but the two are easily distinguished by the latter's long antennae and slender body.

Fig. 4-6
Tumblebug (*Canthon imitator*) rolling cattle dung

Tumblebug
Canthon imitator
(FIG. 4-6)

Biology: Tumblebugs are black beetles that roll rounded balls of dung by using their back legs. They cooperate in pairs to roll the resource some distance from its point of origin before digging a hole to bury it. One partner works at the leading end of the ball, the other at the end that trails behind. Moving the dung away from the pat reduces the danger that it will be stolen or that the adults, and later the young, will be killed by other animals that are roiling about in the complex ecosystem that the pat has become. When the manure is safely stowed within a burrow, the female lays an egg inside it, and the burrow is covered over. The grub upon hatching consumes the dung.

Distribution: This animal has an unusually large range, occurring from as far west as Arizona and Colorado to an eastern limit in Texas, or perhaps as far east as Florida.

Remarks: Length = 16 mm. Drab black tumblebugs are much more common than the glittering green rainbow scarabs, and sometimes one cowpat will host a dozen or more individuals alongside no more than one or two rainbow beetles. The "imitating" dung beetle differs not only by its black coloration but by its curious habit of rolling a dung ball with its hind legs. If and when the rainbow scarab does roll dung, it uses the front of the body instead.

Similar species: It is fortunate that this species is so large, because only one other close relative is likely to be confused with it. That species is the

Fig. 4-7 Delta scarab (*Trigonopeltastes delta*)

Fig. 4-8 Spotted grapevine beetle (*Pelidnota punctata*); mating pair at black light

smooth tumblebug (*C. laevis;* Robinson 1948). The scientific name "imitator" might have been coined because of a great resemblance between the newly discovered species and the other (Brown 1946). The imitator is also distinguished by habitat, having a greater preference for sand, and it replaces its relative entirely on these soils. Thus, it is no surprise to find the imitator species in abundance on the Carrizo Formation of the Ottine wetlands.

Delta scarab
Trigonopeltastes delta
(FIG. 4-7)

Biology: We encountered several of these small, husky, and very pretty adult beetles as they flew from frostweed plants in the understory of Palmetto State Park. Elsewhere they have been observed feeding on wild carrot flowers (*Daucus carota;* Lago and Mann 1987). The grubs develop inside old oak stumps (Dozier 1920).

Distribution: From the Atlantic Ocean to Texas.

Remarks: Length = 9 mm. Adults appear briefly in spring, but the species appears to be uncommon in these wetlands.

Similar species: None. The light-colored triangle (delta) pattern on the dark thorax immediately distinguishes this scarab from all others in the area.

Spotted grapevine beetle
Pelidnota punctata
(FIG. 4-8)

Biology: The adults eat grape leaves and become pests in the eastern United States where the plants are grown commercially. They also feed on Virginia creeper, a close relative of the grape. Though we never saw adults during the daylight hours here or in the Lost Pines forest, they are said to fly during the day from vine to vine, making a loud buzzing sound. Grubs are not harmful because they live in stumps and decaying roots of oak, hackberry, and elm.

Distribution: From the Atlantic Ocean to a western limit along the Rio Grande near Del Rio, Texas.

Remarks: Length = 25 mm. Living specimens should be handled with care because the sharp claws can pierce skin like fishhooks. It isn't clear how they came to be sharper than those of related beetles.

Similar species: None.

Loving scarabs
Phileurus valgus, P. truncatus
(FIG. 4-9)

Biology: Very little has been published about the biologies of these rare and rather exotic scarab beetles beyond the fact that *P. valgus* has been found beneath the bark of decomposing trees (Blatchley 1910). We cultured eight grubs of this species to the adult stage after collecting them from a blackjack oak snag in the Lost Pines forest in July. This is the most common species in the area. Both sexes of both species can exceed 25 mm in length, and each individual has three small horns, two projecting from the head and one just behind. They probably make noises by rubbing certain body parts together, because they bear structures known to have this function in other beetles. We seldom saw either species under natural conditions. A few specimens flew to our black lights and to incandescent lights after dark.

Distribution: Both loving scarabs occur between the Atlantic Ocean and a western limit near the Ottine swamps.

Remarks: A typical length for *P. valgus* = 20 mm; that for *P. truncatus* = 32 mm. The loving scarabs are so poorly known that they do not appear in

Fig. 4-9
Loving scarab
(*Phileurus valgus*)

the various modern books dedicated to the beetles of the United States, and we coined the common name ourselves from a translation of the scientific name. *Phileurus valgus* is usually smaller than its close relative and has more obvious lines on its wing covers. We understand from Internet sources that, quite remarkably, adults of one or both of these species are predaceous upon other insects, even species larger than themselves.
Similar species: None.

Ox beetle
Strategus antaeus
(FIG. 4-10)

Biology: Big, tanklike, mahogany brown adults may approach 50 mm in length and reportedly hide beneath fallen, decaying logs during the day. We believe the grubs of these wetlands live inside the logs and eat the wood itself. We found one adult female deep within a rotting pecan log where she probably developed or was laying eggs. In the same niche we also saw several grubs large enough to be this species but too large to be much else with the exception of the eastern Hercules beetle (*Dynastes tityus*). Yet we saw no adults of the latter and do not know if it occurs in these wetlands. The ox beetle grubs manufacture their own habitation chambers inside the logs as a result of chewing the surrounding wood. Most adults we saw were individuals of both sexes that flew to our black lights at night.

Distribution: From the Atlantic Ocean to a western limit near the Ottine swamps.

Fig. 4-10
Ox beetle (*Strategus antaeus*)

Remarks: Length = 31 mm. Males have three enormous, exotic-looking horns that rise from a plate just behind the head. Females have just one small horn. When these monsters fly into a plastic light trap at night, the crash can be heard from one hundred feet away.

An early study in the eastern United States detailed the larval life cycle as it occurs *underground.* The adults make the nests and show much parental care prior to egg laying (Manee 1908). In the Ottine swamps we know of the appearance of adults and juveniles inside pecan logs where presumably little if any parental care was invested beyond the laying of an egg.

Similar species: None.

Hide beetles
Trox robinsoni, T. suberosus, T. rubricans
(FIG. 4-11)

Biology: Hide beetles are drab, odd-looking scarabs that develop as grubs within the dried remains or sometimes the excrement of other animals. Adults resemble crumbs or small chunks of soil, and someone with a sharp eye can find them in hoof parings, bonemeal, owl pellets, wool coats, and on or in dead fish, birds, reptiles, and mammals. They also inhabit the nests and burrows of warm-blooded animals. Some hide beetles are known only from the nests of a single species of bird, such as the barn owl (Vaurie 1955).

Fig. 4-11
Hide beetle from raccoon carcass (*Trox rubricans*)

The large hide beetle *T. suberosus* (length = 17 mm) has been found in goat carcasses, beneath chicken feathers, "at malt (germinated grain) under cow chips," and at lights after dark. The other two species featured here are a little smaller. In Texas Robinson's hide beetle is known from chicken feathers; a mouse's nest beneath a stone; a light trap at night; and the carcasses of squirrels, goats, and sheep. We found *T. rubricans* (Fig. 4-11) at the dried remains of a raccoon carcass in North Soefje Marsh.

Distribution: Robinson's hide beetle has an unusual distribution. In the United States it occurs only within a central corridor from South Dakota to a southern extreme in Texas near the Ottine swamp (Vaurie 1955). Thus, it is neither western nor eastern in its affinities. Its larger relative *T. suberosus* has one of the widest distributions among the animals treated here, occurring from coast to coast within the United States and elsewhere in more southern parts of the New World as far from the Ottine swamp as Argentina. *Trox rubricans* is a Neotropical species that is rare within the United States except in southern Texas, where we found it.

Remarks: These hide beetles are good fliers and came to our blacklight traps at night with the exception of the individual from a raccoon carcass. Some species have wing covers that look normal, but upon moving them aside one finds small and presumably functionless wings. The Texas hide beetle (*T. texanus*) of the southern part of the state is one of these. It might occur in the swamp, though its currently known northern limit lies several counties to the south.

Similar species: Dozens of hide beetles occur in Texas, which is clearly a center of diversity for the group. Reliable identification requires a difficult formal key, at least a hand lens for magnification, and sometimes not only the male sex but dissection of the same (Vaurie 1955).

Fig. 4-12
Green June beetle
(*Cotinis nitida*)

Green June beetle
Cotinis nitida
(FIG. 4-12)

Biology: All active life-cycle stages feed on living plants. The attractive adults are pests aboveground when they eat fruits and flowers, whereas grubs destroy turfgrass roots underground and out of sight. They do this in a remarkable fashion by crawling on their backs instead of on their legs. In early September we saw hundreds of adults flying into and out of a lone loblolly pine just outside Soefje Swamp, but we have no explanation for this behavior except as a mating congregation. The pine is not native to the wetlands.

Distribution: From the Atlantic Ocean to a western limit near the Ottine wetlands.

Remarks: Length = 23 mm. The metallic green and bronze adults fly mostly at night. When flying during the day, they look and sound like bees. Another common name is fig-eater.

Similar species: None. The smaller, brighter metallic green flower scarab (*Trichiotinus lunulatus;* length = 10 mm; Fig. 4-13) is more likely to be seen

Fig. 4-13
Metallic green flower scarab (*Trichiotinus lunulatus*)

Fig. 4-14
The dark flower scarab (*Euphoria sepulcralis*)

feeding on pollen, as is the dark flower scarab (*Euphoria sepulcralis;* length = 13 mm; Fig. 4-14).

Fig. 4-15
Carrot beetles
(*Bothynus gibbosus*)

Carrot beetle
Bothynus gibbosus
(FIG. 4-15)

Biology: Beyond wild places the larvae of this scarab beetle, known as white grubs, are pests because they eat the roots of carrots and other produce. Adults eat foliage instead (White 1983). Dozens flew to our black lights in May, but we never noticed adults or juveniles in natural settings.

Distribution: From coast to coast within the United States.

Remarks: Length = 16 mm. There is disagreement about the proper scientific name of this animal. Some call it *Ligyrus gibbosus* (Endrödi 1985).

Similar species: Several related species are likely to occur in the wetlands. One must consult keys to distinguish them (Cartwright 1959). They resemble May beetles but are more closely related to the giant ox beetles and Hercules beetles, as indicated by the tiny, hornlike structures on the head of both sexes.

Haldeman's ironclad beetle
Zopherus haldemani
(FIG. 4-16)

Biology: The biology is essentially unknown. We, as well as others, have collected adults on dead oak trees. Adults, grubs, and pupae were once found on a pecan, an abundant tree in the wetlands. We saw several mating pairs on an American elm in the upland bordering Rutledge Swamp. Fire

Fig. 4-16
Haldeman's ironclad beetle (*Zopherus haldemani*); mating pair

ants were crawling on them, but the beetles' extraordinarily tough armor protected them from stings. However, the female ironclads seemed to be laying eggs in crevices, and perhaps the ants were successfully gathering these.

Distribution: This is a Texas species found nowhere else in the United States but in the central and eastern part of the state between Kerrville and Houston, with one outlying record farther west (Triplehorn 1972). Because it also occurs in Mexico, it cannot be described as a Texas endemic. This limited range is shared by several grasshoppers, mantids, and cockroaches and suggests a history of grassland expansions and contractions associated with ice ages and the evolution of the Great Plains.

Remarks: Length = 29 mm. The ironclad is named for an exoskeleton so tough that entomologists sometimes drill a small hole in a wing cover to accept an insect pin that otherwise might bend or break but never penetrate. This raises a second interesting fact. Though the ironclad has wing covers as seen in other beetles, it has no wings beneath them to protect, and the covers are locked shut. In 1976 this beetle was chosen as the official mascot of the Southwestern Entomological Society, and it has appeared on the covers of its journal ever since. It is also known as *Zopherus nodulosus haldemani.*

Similar species: No other animal in or bordering the wetlands has a color pattern resembling the ink blots of a Rorschach personality test.

Fig. 4-17
False mealworm beetle (*Alobates pensylvanica*)

False mealworm beetle
Alobates pensylvanica
(FIG. 4-17)

Biology: Adults are nocturnal omnivores that feed on other insects and fungi. During the day and when overwintering in groups, they hide beneath the bark of oaks and ashes. If handled, they secrete a defensive odor that we found pleasant rather than noxious. They also draw in their legs and feign death. The grub stage lives within decaying logs and feeds upon other insects.

Distribution: From coast to coast within forested regions of the United States.

Remarks: Length = 23 mm. This is no pest, though the true mealworm beetle is (see next section).

Similar species: None.

Yellow mealworm beetle
Tenebrio molitor
(FIG. 4-18)

Biology: Yellow mealworm beetles are named for their larval stage, a grub that achieves pest status because it eats stored grains. We saw the adults instead. We found numerous individuals feeding on glands of coffee senna (*Senna occidentalis*) at the edge of North Soefje Marsh. It is likely that

Fig. 4-18
Yellow mealworm
beetle (*Tenebrio
molitor*) feeding on
coffee senna (*Senna
occidentalis*)

these nonnative beetles were introduced into the Ottine wetlands via containers of infested livestock feeds.

Distribution: Yellow mealworms are found throughout the world where
grains are imported and stored and where pets are fed with the larval
stage.

Remarks: Length = 16 mm. This widely distributed pest can be described as
a zoological weed. The beetle's grub stage is the ubiquitous mealworm
raised in pet stores to feed reptiles and amphibians.

Similar species: Dark mealworm beetles (*T. obscura*) might occur in the wetlands. They too are widespread pests. Adults may be distinguished from
the yellow mealworm beetle by their darker color and by the nonreflective
surface of the body.

Giant rove beetle
Platydracus maculosus
(FIG. 4-19)

Biology: Our single encounter with this elongate, blackish beetle occurred in
a pasture near Rutledge Creek. We found one mating pair in mid-June on
a cowpat where dung beetles were digging underground and rolling
manure to safer sites. The rove beetles were there, not to collect dung
themselves but to kill and eat the scavengers that were doing so. Even
large and heavily armored dung-rollers can be captured, killed, and consumed within fifteen minutes (Young 1982).

Fig. 4-19
Giant rove beetle
(*Platydracus maculosus*)

Distribution: These "flat dragons" occur from the Atlantic Ocean to a western limit that remains undetermined.

Remarks: Length = 25 mm. This is the largest, or one of the largest, rove beetles in the wetlands. It is known by two other names: *P. viduatus* and *Staphylinus maculosus.*

Similar species: None that we saw, though perhaps half a dozen might occur in the area.

Bumelia borer
Plinthocoelium suaveolens
(FIG. 4-20)

Biology: We saw this gorgeous, long-horned, wood-boring beetle while exploring upland adjacent to Rutledge Swamp. Its colors are metallic green, orange, and black. Several adult females and one male were collected as they left their host plant, the ironwood tree (*Sideroxylon lanuginosum*). This particular plant was alive but damaged, leaning almost horizontally, and was presumably more susceptible to attack than a healthy tree. The beetles crawled rapidly near ground level, and one was taken by net in midair as it attempted to fly away. The grub stage bores within the roots of bumelia, and when grubs occur in numbers, they kill the tree (Linsley and Hurd 1959). Adults visit flowers for nectar and/or pollen.

Distribution: From the Atlantic Ocean to Arizona.

Remarks: Length = 35 mm. Bumelia borers have been reported to develop within mulberry trees and tupelo (Hamilton 1892; Linsley 1964). The

Fig. 4-20
Bumelia borer
(*Plinthocoelium suaveolens*)

former occurs in the Ottine swamp, but the latter does not. Their bumelia hosts are more characteristic of the border between wetland and upland than of the wetland itself.

Similar species: None.

Giant stump borer
Stenodontes dasytomus
(FIG. 4-21)

Biology: This is a huge, brown beetle. It is equipped with powerful jaws and an appearance that seems more at home in a tropical forest than in a Texas wetland. Trees attacked by its grub stage include oaks, willows, hackberries, sycamores, boxelder, and pecan (Duffy 1960; Linsley 1962a). They bore within the wood of living, weak, and dead trees for up to four years and do much damage in the process because of their large size. The pupal stage is spent within the tree rather than in the soil, and in Texas escaping adults chew their way through wood and bark from April to August. These holes may be seen near the base of tree trunks along the Palmetto Trail. In Arizona the damage to sycamores was primarily in areas already withered by sunscald, especially on the west side of trees (Linsley, Knull, and Statham 1961).

Adults are attracted to lights, but we had an interesting experience with one individual that seemed to contradict this well-known behavior. In mid-April we encountered it, a male and the largest specimen of all, peer-

Fig. 4-21
Giant stump borer (*Stenodontes dasytomus*); mating pair at black light

ing out from a hole in the fabricated sandstone wall of a water tower. In the flashlight's beam the beetle suggested a tiny dragon staring out from the entrance to its lair. When the light was turned off, the beetle ventured partly out of the hole but retreated instantly whenever the spotlight was turned in its direction. This behavior somehow suggested a small, intelligent mammal. Perhaps the individual was a newly transformed adult. These are known to remain near the exit hole of their natal tree for some time, and they do hide in the hole during the day. In Arizona, and true to our own observations in south-central Texas, "in the early evening and at night they may be seen with their heads and antennae protruding from large oval exit holes or crawling over the surface nearby" (Linsley, Knull, and Statham 1961, 5; see photo therein). However, our specimen was holed up in a crevice of a stone wall, which is obviously no source of food to the grub stage. Perhaps it developed in long-dead supporting timbers hidden behind the stone.

In the heat of day both sexes sometimes crawl about, and the females back into their own exit holes to lay eggs. Again remarkable is their ability to stridulate audibly and their apparent propensity to do so even when *not* provoked: "They are capable of making a loud clicking sound, which can be heard at a distance of 6 feet, when they are not disturbed" (Linsley, Knull, and Statham 1961, 5).

Giant stump borers are prey for nocturnal birds, bats, great horned owls, and in Arizona, a fellow beetle closely related to the eyed elater. Thus, the eyed elater of the Ottine wetlands presumably preys upon the borer as well. Tiny pseudoscorpions hitchhike beneath the wing covers and can survive freezing temperatures that kill their host.

Distribution: This beetle is regarded as a tropical species that ranges north into the southern United States, where it occurs from the Atlantic Ocean to Arizona.

Remarks: Length = 46 mm. The huge grub is a delicacy in tropical countries. The simple "recipe" calls for soft, white larvae to be roasted slowly over a charcoal fire.

Similar species: None in south-central Texas. Males may be identified by their particularly large and hairy jaws.

Giant red and black oak borer
Leptura gigas
(FIG. 4-22)

Biology: Adults are large, attractive, long-horned beetles that resemble spider wasps in size, color, and even in the manner of their flight. They presumably eat wildflower pollen if they feed at all, but this is one species that seems to stay within or near the timber. The grub stage feeds in black willow, which was abundant where we found the adults.

Distribution: Nearly endemic to Texas, though also occurring just across the Rio Grande in Mexico.

Remarks: Length = 31 mm. We saw only a handful of these impressive animals. Some flew to black lights at night, but one, perhaps the closely related species *L. emarginata,* was caught among piles of cut oak logs. Perhaps it had developed inside a log or had been attracted to the wood to lay eggs.

Fig. 4-22
Giant red and black oak borer (*Leptura gigas*)

Similar species: Only one beetle resembles this one enough in both color and size to cause confusion. The red and black oak borer *L. emarginata* is an eastern species that reaches its western limit in central Texas near the Ottine swamps. Its wing covers are smooth rather than ribbed. If others wish to attract these two rarely encountered beetles, they might try applying a concoction of fermenting syrup bait to the bark of trees, which we used to great effect in the nearby Lost Pines forest. Confident separation of the two sister species requires close examination and the proper identification keys.

A much smaller and similarly scarce beetle resembles *L. emarginata* in shape and color. We found a single specimen of this species, the little hickory borer (*Trigonarthris proxima;* length = 12 mm), on Drummond's rough-leaved dogwood (*Cornus drummondii*). It has been previously reported from dogwood elsewhere. Adults also fly to black lights, though we never saw one at ours. The grub stage feeds within decaying hardwoods, including hickory (Yanega 1996). This attractive little animal occurs from the Atlantic Ocean to a western limit somewhere in Texas, perhaps in or near the Ottine wetlands.

Cottonwood borer
Plectrodera scalator
(FIGS. 4-23–4-24)

Biology: This black and white, long-horned, wood-boring beetle is perhaps the most attractive longhorn in Texas and is certainly among the largest. The grub stage feeds and burrows within cottonwood trees and black willows, where it can cause great damage over a two-year period, including a feeding behavior that may girdle the tree (Linsley and Chemsak 1984).

Adults have black wings and do fly, though we have never seen what must be an impressive flight. Males fight aggressively, and the winner is the beetle that seizes its opponent's antenna first (Goldsmith et al. 1996). Females lay their eggs on the trunk at or a little below ground level (Fig. 4-23). When handled, the struggling insect often stridulates, making chirping noises by rubbing plates of its thorax together.

Distribution: From the Atlantic Ocean to New Mexico, with perhaps its greatest concentration within the central plains states.

Remarks: Length = 40 mm. Cottonwood borers sometimes erupt in large numbers from trees growing well within city limits in central Texas. Biological control efforts with fungi have been successful (Forschler and Nordin 1989).

Similar species: None.

Fig. 4-23
Cottonwood borer (*Plectrodera scalator*) at base of enormous host tree at Rutledge Creek

Fig. 4-24
Cottonwood borer with black wing exposed

Fig. 4-25
Banded hickory borer (*Knulliana cincta*)

Banded hickory borer
Knulliana cincta
(FIG. 4-25)

Biology: Males and females arrived at our black lights in spring. This is a large, long-horned wood borer, and the antennae of the male are much longer than its body. Females lay their eggs in dead branches where the grubs burrow in wood, creating lengthy galleries before pupating in the tree rather than in the soil. Banded hickory borers use a wide variety of host plants, including pecan, elm, mesquite, oak, and willow (Linsley 1961; Linsley and Chemsak 1997).

Distribution: From the Atlantic Ocean to west Texas with a disjunct range in southern Arizona.

Remarks: Length = 32 mm. Our first sighting of this conspicuous species did not occur until after we had blacklighted all four seasons of an entire year. The diversity of long-horned wood borers in central Texas is daunting, and previously unseen species are likely to occur with each new outing.

Similar species: None.

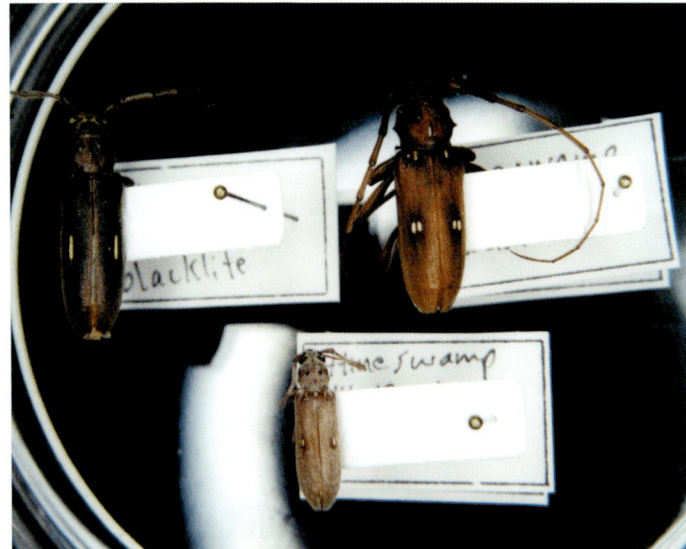

Fig. 4-26
Dark ivory-marked
beetle (*Eburia ovicollis;* top left),
Haldeman's ivory-marked beetle
(*Eburia haldemani;*
top right), and the
lesser ivory-marked
beetle (*Eburia
mutica;* bottom)

Ivory-marked beetles
Eburia ovicollis, E. haldemani, E. mutica
(FIG. 4-26)

Biology: Ivory-marked beetles are long-horned wood borers readily distinguished by the small ivory-colored marks that stand out against the background of their much darker wing covers. In south-central Texas the grub stages feed on wood or bark of mesquite (the dark ivory-marked beetle, *E. ovicollis*); rough hackberry, mulberry, and willow (Haldeman's ivory-marked beetle, *E. haldemani*); and in mesquite, smooth hackberry, elm, Herculesclub, sesbania, and boxelder (the lesser ivory-marked beetle, *E. mutica;* Szafranski 2002).

Distribution: The dark ivory-marked beetle and the lesser ivory-marked beetle occur from central Texas to northern Mexico, whereas Haldeman's ivory-colored beetle has a much greater range, occurring from the Atlantic Ocean to New Mexico (Linsley 1962b).

Remarks: Length of *E. ovicollis* = 22 mm; length of *E. haldemani* = 26 mm; length of *E. mutica* = 21 mm. We saw these three species only at black lights after dark. In August we observed all but the lesser ivory-marked borer while they mated there.

Similar species: Other ivory-marked wood borers might occur in the swamp. Identification requires formal keys. One relative, the patriarchal grub, is reported to live more than forty years (Arnett, Downie, and Jaques 1980).

Fig. 4-27
Red oak borer
(*Enaphalodes rufulus*)

Red oak borer
Enaphalodes rufulus
(FIG. 4-27)

Biology: Grubs burrow into the tree after the female lays eggs on the bark of oaks and maples (Linsley 1963), presumably including the boxelders that are so abundant in the Ottine swamps. When the feeding period ends, the grub cuts a hole to the outside for its eventual emergence as an adult and then retreats into the heartwood to pupate behind the protection of a plug of its own waste. Populations can increase to the point of killing the host, and the red oak borer becomes a pest when it kills shade trees within city limits.

Distribution: From the Atlantic Ocean to west Texas.

Remarks: Length = 28 mm. A living grub was found inside a twenty-three-year-old oak table in Britain (Hickin 1958). It is not certain what to make of this, for the female is said to use living trees instead of dead ones, suggesting that the grub was more than two decades old. Because the report does not mention raising the grub to adulthood before identification and because identification is difficult, the remarkable report is hard to accept.

Similar species: None. A much darker and smaller species that also appears at black lights is the black oak borer (*Parelaphidion aspersum*), sometimes seen in mating pairs in August. It has been reported only from pignut hickory (*Carya glabra*) and swamp chestnut oak (*Quercus prinus*), neither of which occurs in the swamp, so a new host record awaits discovery. Perhaps it is the mulberry tree fed upon by the grub of a closely related species. Similar to the red oak borer in size but blackish like the above-mentioned species is the willow borer (*E. taeniatus*). In the United States this species is restricted to central and mainly south-central Texas, with at least one record in northern Mexico in the Matamoros region (Linsley 1963).

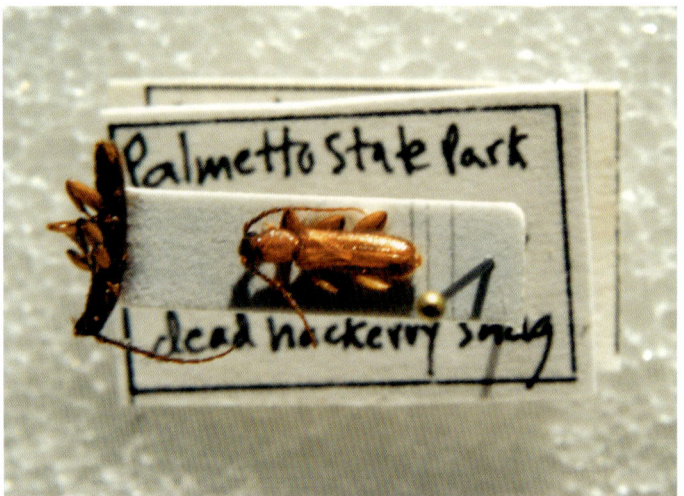

Fig. 4-28
Flat powder-post beetles (*Smodicum cucujiforme*)

Flat powder-post beetle
Smodicum cucujiforme
(FIG. 4-28)

Biology: Grubs develop in oaks, willows, and hackberries after they hatch from eggs laid in the crevices of exposed wood (Linsley 1962b). They tunnel in very old, dry, seasoned heartwood and pupate without leaving the tree. We found the small, extraordinarily flat adults beneath the bark of decaying hackberry snags.

Distribution: From the Atlantic Ocean to west-central Texas.

Remarks: Length = 8 mm. The species name *Smodicum texanum* has appeared at least once in the literature, though we are not sure of its validity (Linsley and Chemsak 1997). In any event it probably does not occur here because its only known host is the great lead-tree (*Leucaena* *pulverulenta*) that, within Texas, is limited to the extreme south of the state.

Similar species: The banded wood borer (*Obrium maculatum*) might be found on the same host trees, but it has dark brown bands on its wing covers.

Fig. 4-29
Dogwood borer
(*Oberea tripunctata*)

Dogwood borer
Oberea tripunctata
(FIG. 4-29)

Biology: We caught a single specimen flying near a dogwood tree in Rutledge Swamp. The adult female girdles living twigs of dogwoods, elms, and even a few herbaceous plants, and upon hatching from eggs the grubs mine within the wood or stem (Ruggles 1915). Pupation occurs within the plant rather than in the ground (Craighead 1923).

Distribution: From the Atlantic Ocean to Texas.

Remarks: Length = 16 mm. Some report that the larvae are very active and move rapidly, though we saw no juvenile stages.

Similar species: Several close relatives of the dogwood borer occur in central Texas. Formal keys are required for confident identification.

Fig. 4-30
False longhorn bee-
tles (*Distenia
undata*)

False longhorn beetle
Distenia undata
(FIG. 4-30)

Biology: Two adults flew to our black light in early May. Others have seen
them, rarely, on the foliage of wild grape and running about on bark at
night, when females lay eggs at the base of trunks of dead host trees.
These include oak, hickory, elm, redbud, and, surprisingly if true, pine
(Yanega 1996). If disturbed, the adults are likely to drop to the ground
and feign death. At such times they become difficult to detect (Blatchley
1910) because of disruptive coloration that blends into the background,
much as tiger stripes do. Grubs feed in the wood of the host tree and
eventually pupate within the tree rather than in the ground. Development
from egg to adult probably requires two years.

Distribution: Unknown. The false longhorn appears to be an eastern species
occurring from the Atlantic Ocean to a western limit somewhere in Texas.

Remarks: Length = 26 mm. There is agreement from literature and personal
communication that this is a rarely seen species. We saw it only once.
Those interested in the enigma that it presents might wish to consult a
difficult-to-find work on an Asian relative that has been more closely
studied, because its biology is likely to be quite similar to that of our
species and because illustrations of adult, larva, and pupa are provided.
This species, *D. gracilis,* is only sporadically encountered. Its grub stage
lives below ground in the wood of roots of dead trees, and when development is completed, the adult chews its way to freedom (Cherepanov
1988).

Similar species: None. This beetle we describe as the false longhorn because it lacks an anatomical trait that, according to some, does not allow it to belong to the family of long-horned, wood-boring beetles, despite its overall appearance. We thank Steven W. Lingafelter of the United States National Museum for providing us with information on this intriguing animal from published sources unavailable to us at the time.

Fig. 4-31
Hackberry long-horned borer (*Urographis triangulifer*)

Hackberry long-horned borer
Urographis triangulifer
(FIG. 4-31)

Biology: Little is known of this creature's habits. The female has a lengthy ovipositor that she uses to lay eggs within the trunks of hackberries and maples, presumably including boxelder. All of these trees are abundant in the Ottine swamps.

Distribution: From Alabama to a western limit somewhere in Texas.

Remarks: Length = 17 mm. Hackberry long-horned borers are known by a number of scientific names. We follow the most recent treatment (Linsley and Chemsak 1995).

Similar species: The centered dark patch at the base of the wing covers distinguishes the hackberry long-horned borer from others that resemble it, such as the boxelder borer (*Sternidius variegatus*), which has a distinctive white half-ellipsoid near the tip of each wing cover in combination with attractively striped antennae and banded legs.

Fig. 4-32
Texas ash borer
(*Anelaphus
niveivestitus*)

Texas ash borer
Anelaphus niveivestitus
(FIG. 4-32)

Biology: The Texas ash borer is a small long-horned beetle that feeds inside ash and hackberry trees during its grub stage. Little else is known about its biology except that specimens were once beaten from ash trees in a palmetto grove in south Texas (Linsley and Martin 1933). Adults lie motionless on the collecting sheet and are apt to be overlooked. Texas ash borers were previously reported to fly only from May to June, but two individuals flew to one of our black light traps in late April and mated on the spot.

Distribution: This is a Texas-endemic species currently known only from the southern and southeastern part of the state.

Remarks: Length = 12 mm. The Texas ash borer was discovered at Esperanza Ranch in the Brownsville area of south Texas and made known to science in 1905.

Similar species: Other small reddish brown longhorns are likely to be confused with this species, including the oak twig-pruner (*Elaphidionoides villosus;* length = 18 mm), which is larger than the endemic Texas ash borer but similar enough to require the relevant key and at least the magnification power of a hand lens (Linsley 1963).

Fig. 4-33
Vogt's hackberry
borer (*Neoclytus
mucronatus vogti*)

Vogt's hackberry borer
Neoclytus mucronatus vogti
(FIG. 4-33)

Biology: This is a wasplike, long-horned wood borer that develops in the
grub stage within a variety of trees growing in and near the swamp. These
include false buckthorn (*Sideroxylon lanuginosum*), smooth hackberry (*Celtis
laevigata*), Jerusalem thorn (*Parkinsonia aculeata*), mesquite (*Prosopis glandulosa*), and cedar elm (*Ulmus crassifolia;* Linsley 1964). They are reported to
prefer dead and dying plants, but we collected several specimens from
seemingly healthy trees.

Distribution: From Texas to Arizona.

Remarks: Length = 23 mm. This is one of the relatively few western insects
occurring in the Ottine swamp. A different subspecies uses hickory as a
host in place of those used by Vogt's subspecies.

Similar species: The redheaded ash borer is smaller in size, and the yellow
bands on its back constitute a different color pattern.

Fig. 4-34
Redheaded ash borer
(*Neoclytus acuminatus*)

Redheaded ash borer
Neoclytus acuminatus
(FIG. 4-34)

Biology: The redheaded ashborer is a fast-running, long-horned beetle that resembles a stinging wasp so much in color, shape, and behavior and in its long legs that professionals have hesitated before collecting it by hand (Townsend 1884). The larval stage feeds within the wood of a great variety of trees, including ash, oak, hickory, and persimmon (Linsley 1964). Adult females, which may be on the wing for most of the year in south Texas, are choosy about the condition of the dead tree, for they lay eggs only in unseasoned wood with its bark still attached. The small size of the borer is misleading because its numbers may grow large enough to riddle sapwood until millions of feet of ash logs are destroyed. These grubs remain inside the wood until they transform to the adult stage.

Distribution: From the Atlantic Ocean to Colorado.

Remarks: We discovered many adults that were running rapidly across a fallen log lying only a few feet from the San Marcos River.

Similar species: This is the smallest species of its general shape and color that we noticed. Though it may be as large as 12 mm, it may also be as incredibly small as 4 mm.

Fig. 4-35
Patent leather beetle (*Odontotaenius disjunctus*) with heavy load of mites

Patent leather beetle
Odontotaenius disjunctus
(FIG. 4-35)

Biology: The patent leather beetle's life history is remarkable because this very large species lives in loose societies within pecan logs and, to a lesser extent, logs of other trees. All stages of the life cycle are present: eggs, grubs, pupae, and adults. Members of the society communicate by making vibrations that we perceive as sound. The adults generate creaking, hissing noises when they rub the abdomen against overlying wings that in turn lie beneath protective covers. The covers seldom open because this beetle rarely uses its large, well-developed wings to fly and usually walks wherever it goes. Grubs are, of course, forced to use a different set of organs for making sounds. The third pair of legs is reduced to a pair of stumps that scratch against the more typical legs just in front. These stumps are shaped like tiny bear paws and produce their sounds by raking up and down (Reyes-Castillo and Jarman 1980). Stridulation also deters predation by crows (Buchler, Wright, and Brown 1981).

Female patent leather beetles will carry their eggs in their jaws while searching for a suitable place to set them down if human investigators create unnatural conditions in the laboratory. Both sexes feed the grubs indirectly by plastering tunnels inside the log with chewed-up pieces of wood. Grubs eat the pulp as they crawl about, and both they and the adults eat termites and cannibalize injured juveniles.

When the grub is ready to become an adult, it makes a depression in the soft wood of a tunnel floor by rolling back and forth like a buffalo in its wallow. Then it crawls into the depression, and the adults cover them over with wood and/or mud. This protects the soft, vulnerable pupa from being trampled or eaten by other members of the society. Grubs will accept a hollow made by the pressure of a human thumb (Gray 1946). When metamorphosis is complete, the insect appears in beautiful but temporary colors of orange and white that soon change to shiny black as the exoskeleton hardens. Every adult seems to harbor numerous mites that crawl about on the undersurface and cling to the mouth area, sometimes appearing on top en masse.

Distribution: Patent leather beetles range from the Atlantic Ocean to a western limit near the Ottine swamps.

Remarks: Length = 36 mm. Dispersal from one log to another presumably occurs by overland march, because the wings are seldom used for flight. In fact, there are only two reports of flight in this species. One was a mating flight with both sexes united in the air (Macgown and Macgown 1996).

Midwestern children at the beginning of the twentieth century amused themselves with the beetle known to them as "the horn," in reference to the small tubercle that all adults bear on the head. They tied one end of a string to the horn and the other end to some object for the beast of burden to tow. Experiments later showed that patent leather beetles can haul seven pounds, or three thousand times their own weight (Hinds 1901). Other common names for the patent leather beetle are pegbug, Betsy beetle, Bess beetle, and horned passalus.

Similar species: None.

American oil beetle
Meloë americanus
(FIG. 4-36)

Biology: Oil beetles are named for an oily, yellowish, defensive liquid that oozes from the joints of the disturbed animal's legs. This fluid and/or other parts of the body contain the chemical cantharidin, which is variously used as an aphrodisiac (the so-called Spanish fly), as a healing drug, or, at the other extreme, as a poison. Seldom seen or collected, these oil beetles are perhaps even more remarkable for their anatomy, behavior, and life cycle. For example, adults cannot fly because they have no hind wings. Their forewings are also highly reduced. When disturbed, they call upon other defenses in addition to or instead of bleeding at the joints. They feign death by falling on one side, they regurgitate, they kick, and they charge when the opponent is of similar stature (Pinto and Selander 1970).

Adult oil beetles are herbivores that eat flowers, stems, and leaves, and in the Ottine swamp the favored foods are likely to include all three but-

Fig. 4-36
American oil beetle
(*Meloë americanus*)

tercup species. The grub is a part-time herbivore at best. Juvenile oil beetles in general are predators, parasites, or combinations of these depending
upon the species and also eat plant food in the form of their host's nest
provisions. One of many stages in the development of the grub is known
as the "triungulin." It hatches from the egg, climbs into flowers, and
hitches a ride on the fur of bees that arrive to collect pollen and nectar
but depart with a natural enemy onboard. When the bee returns to its
nest, which in the case of the American oil beetle is perhaps the nest of a
sweat bee, the oil beetle grub dismounts and begins eating the provisions
and/or the grubs of its host. Foraging bees can become so heavily infested
that they cannot fly. One individual had eighty-one triungulins clinging to
its hairs. The fate of grubs that attach to flower-visiting scarab beetles is
unknown. They probably die for lack of the proper host and food.

We saw several soft, 25 mm long, pear-shaped, black and metallic blue
females crawling at night on a trail only a few feet from the swamp. We
observed another female digging a nest in the soil for her eggs (Fig. 4-36)
and a lone male as well. One individual crawled to a streetlight long after
dark, and another was found overwintering on a cactus pad. The time to
see them is late March, for they quickly disappear for the remainder of
the year.

Distribution: American oil beetles occur from the Atlantic Ocean to a western limit in central Texas near Kerrville. We never saw this species during
years of exploration in the nearby Lost Pines forest.

Remarks: Length = 24 mm. If it seems strange for oil beetles to lose their
wings and therefore the ability to disperse to new sources of food by

flight, perhaps this role has been taken over by the juveniles that hitch a ride on the wings of a different species.

Our discovery of the American oil beetle in the Ottine swamp is significant because according to the authority on these animals, "Texas is not *Meloë* country," and "considering the rarity of species other than *laevis* in Texas, whatever you collected would be of some interest" (John Pinto, pers. comm., 2001). The American oil beetle is not as well studied as some of the other species occurring in the United States, and the Ottine swamp would be a fine place to do the necessary fieldwork.

Similar species: Other oil beetles might occur in the swamp. As of this writing, the authoritative guide to oil beetle identification and biology is the work by Pinto and Selander (1970).

Caterpillar hunters
Calosoma scrutator, C. wilcoxi, C. alternans sayi, C. macrum
(FIGS. 4-37–4-40)

Biology: The largest, most attractive, and by far the most abundant of the caterpillar hunters is the fiery searcher (*C. scrutator;* length = 35 mm). Few beetles of the swamp approach its size, and even fewer are so vividly marked in shining colors of metallic blue, green, red, and gold. In both the colorful adult stage and in the black, armored grub stage the fiery searcher is a voracious consumer of pestiferous caterpillars and is thus considered a beneficial insect. Among its prey are the defoliating forest tent caterpillars (*Malacosoma disstria;* Fig. 3-35), themselves clad in a beautiful coat of blue.

Searchers race over the ground and climb into the trees to satisfy a voracious appetite and are capable of consuming, or at least killing, more than one hundred caterpillars over a lifetime. When handled, the adults dispense a foul-smelling chemical more noxious than the odors released by stink bugs, and their scissorlike jaws are capable of delivering a noticeable nip. They can live for three or four years, and at the approach of winter adults hibernate in cells within the soil or beneath stones.

Searcher grubs and adults differ in their prey preferences and in their mode of attack. The smaller, softer, and slower grubs favor helpless, immobile female pupal stages, which they devour until the hapless moth or butterfly is little more than a hollow shell (Gidaspow 1959). Being less capable than the adult beetle, the juvenile's preference is understandable. Female insects are generally more nutritious than males, which also contributes to the grub's preference. When they do attack caterpillars, they strike from below or from the side, and if the prey is protected by hairs, they pinch it in a bare spot between the segments. Adults are more direct.

Fig. 4-37
Fiery searcher (*Calosoma scrutator*) eviscerating May beetle

Fig. 4-38
Wilcox's searcher
(*Calosoma wilcoxi*)

They bite the caterpillar in the middle of the back and kill it quickly regardless of size.

The best way to observe the beautiful fiery searchers is to attract them with a black light trap at night, but their appearance in numbers can only be counted on in the month of April. According to earlier observations by U.S. Department of Agriculture (USDA) workers studying the beneficial behaviors of searchers, "After dark the beetles made a droning noise, probably by moving the wings rapidly, which was audible for quite a distance from the cage" (Burgess and Collins 1917). This was observed under laboratory conditions. We never heard it in the field.

Wilcox's searcher (*C. wilcoxi;* length = 20 mm) would be indistinguishable from the fiery searcher without the aid of a microscope or hand lens

Fig. 4-39
Say's searcher (*Calosoma alternans sayi*)

Fig. 4-40
Smooth searcher (*Calosoma macrum*)

were it not for the smaller size of the former. We found one male crawling up a mesquite tree in broad daylight, one on a trail after dark, and another at a black light on the same April evening. However, this species is not common, and in 2002 a Connecticut government Web page reported the belief that Wilcox's searcher has been extirpated from that state. It, on the other hand, was one of the many enemies (including our own human species) that apparently drove the Rocky Mountain locust to a similar fate in the early years of the twentieth century. At one time that grasshopper was the worst cereal pest in the western United States and appeared with the effect of a biblical plague.

Say's searcher (*C. alternans sayi;* length = 24 mm) and the smooth searcher (*C. macrum;* length = 27 mm) are blackish beetles with little similarity to the fiery searcher and the species named for Wilcox. The beetle named for the early American entomologist Thomas Say is distinguishable from the smooth searcher at some distance by the rows of gold spots on its wing covers. A few observations of its behaviors have been published, these underlining a biological similarity to the other caterpillar hunters, but nothing seems to be known about the biology of the related smooth searcher beyond our own observations in the wild swamp and in the busy heart of the state capital in Austin, Texas. In the swamp it appears in very small numbers at black lights in April, when the fiery searcher is flying to lights by the dozens. In the city it can be found at the same time of year crawling on the limestone wall surrounding the University of Texas campus after dark, moving like a small black shadow in the streetlights' dim illumination.

Distribution: The fiery searcher and Say's searcher occur from coast to coast within the United States; Wilcox's searcher ranges from the Atlantic Ocean to Texas; and the smooth searcher is particularly interesting because within the United States it seems to occur only in Texas and Louisiana and perhaps Arkansas (Gidaspow 1959).

Remarks: In years of study we found only one of these four species in the nearby Lost Pines forest (the fiery searcher). The greater presence of caterpillar hunters in the swamp is perhaps due to the fact that caterpillars themselves seem to be everywhere, at least in spring, and to a lesser extent in summer, whereas they are much less abundant in the dry upland forest. Fragments of the fiery searcher's metallic wing covers can be seen shining in the sun in the scat of raccoons and skunks, which eat the beetles despite their noxious chemical defense.

Similar species: Several additional searchers might occur in the Ottine swamp. In the year 2002 the standard reference was more than forty years old (Gidaspow 1959).

Purple ground beetle
Dicaelus purpuratus
(FIG. 4-41)

Biology: The purple ground beetle is a large, bulky, and coppery purple predatory beetle. It has a specialized diet consisting of snails so plentiful in the Ottine swamp that one can't help crunching them underfoot while walking after dark immediately following the spring rains. Whereas some snail specialists have lengthy jaws that reach inside the opening of the shell, these beetles have short, powerful mandibles, and they use brute force rather than precision to crush the shell. Purple ground beetles spend

Fig. 4-41
Purple ground beetle
(*Dicaelus
purpuratus*)

the day beneath logs and scurry about on the soil after dark.

Distribution: The splendid ground beetle (*D. purpuratus splendidus*) is a sub-species with an unusual distribution, because it is confined to the region between Illinois in the east and Arizona in the west. Thus, it is considered a Great Plains animal. We did not see it in our extensive survey of the Lost Pines forest, though we saw it on our first trip to the swamp and on many trips made six years later.

Remarks: Length = 25 mm. The members of this genus are said to be the most beautiful of all North American ground beetles, and the "splendid" subspecies with its coppery sheen tops the list (Say 1825). "It is not rare but scarce enough to give the collector a real thrill when he finds one" (Arnett, Downie, and Jaques 1980, 89).

According to an authority, the populations occurring between Austin and Kingsville, Texas (for example, those of the Ottine swamps), might be different enough to warrant consideration as a separate subspecies altogether (Ball 1959).

Similar species: None.

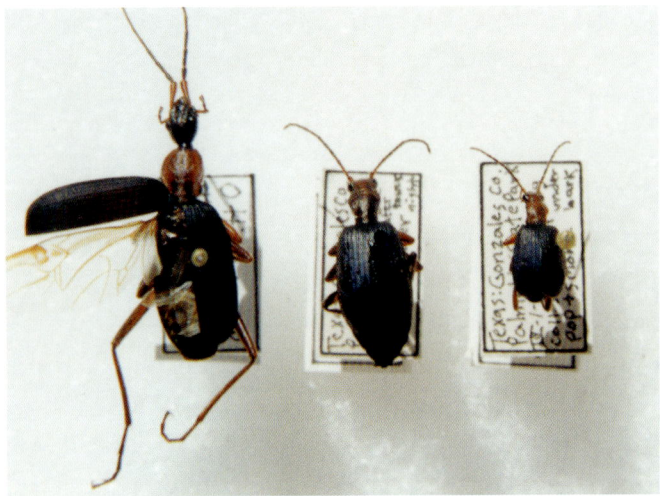

Fig. 4-42
LeConte's false bombardier beetle (*Galerita lecontei lecontei;* left), the large bombardier beetle (*Brachinus alternans;* center), and the small bombardier beetle (*Brachinus sublaevis;* right)

Bombardier beetles
Brachinus alternans, B. sublaevis
(FIG. 4-42)

Biology: The predatory adult bombardier beetle is famous for possessing a defensive apparatus that can be compared to the exhaust pipe of an automobile. This apparatus expels gas from glands at the animal's back end with a popping sound, and the product is visible from some distance as a puff of smoke. The explosion can also be aimed. We saw, heard, felt, and smelled this defense when collecting both species.

Females of at least some bombardier species and perhaps these two as well lay their eggs in mud cells constructed on stems, twigs, or stones (Arnett 1960). The grub or larval stage is a parasite of water beetles such as the whirligigs of central Texas. This explains the abundance of adults along a trail that skirts the lagoons of Palmetto State Park.

We first encountered the larger species, *B. alternans,* on an old sandstone water tower constructed in the 1930s by the Civilian Conservation Corps. The smaller species, *B. sublaevis,* was found initially beneath the bark of a log. Both may be seen running about after dark on trails and on tree trunks. They feed opportunistically upon crushed snails.

Distribution: Both bombardier beetles are eastern animals that occur from the Atlantic Ocean to El Paso, Texas (Erwin 1970).

Remarks: Length of *B. alternans* = 15 mm; length of *B. sublaevis* = 9 mm. The explosive defense of the bombardier beetle is startling but harmless to humans. It does leave a brown stain on the fingers that takes some time to rub off.

Similar species: Many bombardier species occur in southern Texas, and because of their great similarities they are difficult to identify. We thank Terry Erwin of the United States National Museum for helping us identify the Ottine species.

LeConte's false bombardier beetle
Galerita lecontei lecontei
(FIG. 4-42)

Biology: LeConte's false bombardier beetle is a large, primarily predatory species that presumably kills and eats destructive caterpillars, if the biologies of its closest relatives hold true for this species. The larval stage is also expected to be predatory and is very active. We found one grub while we were tearing apart a rotting log in the swamp.

These beetles are known as false bombardier beetles because they bear an outward resemblance to the true variety. They, too, spray a defensive solution to ward off attackers. Collected specimens commonly smell strongly of this secretion, which is 80 percent formic acid (Rossini et al. 1997). To the surprise of the discoverers, the females of a closely related species lay their eggs within small mud cells that they attach to the undersurface of leaves (King 1919). It is not known if LeConte's false bombardier beetle does the same.

The habitat of this subspecies has been described as "damp open ground, in the flood zones of rivers and creeks, or at the margins of marshes" (Ball and Nimmo 1983). We found it in precisely such an environment and usually at night. The beetles can also fly (Forsyth 1991), but we never saw one flying.

Distribution: This subspecies of LeConte's false bombardier beetle has a remarkable range, which is divided into two greatly disjunct sections. The western part extends from the Pacific Ocean to north-central Mexico (Ball and Nimmo 1983). The eastern portion extends from the Atlantic Ocean to coastal Texas with our Ottine swamp record marking a new western limit within this eastern distribution. We never saw LeConte's false bombardier beetle in the more upland Lost Pines forest of central Texas.

Remarks: Length = 21 mm. During daylight hours adults are sometimes found hiding beneath logs and stones. The grub is much faster on its feet than one expects from a larval beetle.

Similar species: There are approximately one-half dozen species in this genus, and they are difficult to distinguish from one another (Ball and Nimmo 1983). False bombardier beetles resemble true bombardier beetles, but the head is dark rather than red. There is an alternative spelling for this genus: *Galeritula.* We thank George Ball for help in identifying the species at hand and for suggesting that we keep an eye out for a huge and possibly extinct ground beetle that has not been reported since the 1940s. *Cyclotrachelus gigas* is known only from Victoria and Kleburg counties, which are near the Ottine swamp.

Fig. 4-43
Pedunculate ground beetle (*Scarites quadriceps*)

Pedunculate ground beetle
Scarites quadriceps
(FIG. 4-43)

Biology: Pedunculate ground beetles are common and attractive insects that may be seen in spring on the ground beneath lights at night. We don't know if they can fly, and the diet, though likely omnivorous, is unknown.

Distribution: From the Atlantic Ocean to Montana.

Remarks: Length reportedly as great as 30 mm, though we have not seen any specimens so large. Considering the abundance of this apparent beetle and the few related species that occur in Texas, it is remarkable that no key to their identification has been published. The literature as of 2002 could not be relied upon for positive identification. We thank George Ball for providing us with an unpublished key to the three Texas species recognized at the time of writing: *S. quadriceps, S. subterraneus,* and *S. lissopterus.*

Similar species: *Scarites subterraneus* (length = 20 mm or less) is a much smaller relative that we believe we have seen in summer when it largely replaces *S. quadriceps* at lights after dark. However, we have not identified these beetles with the newly available key.

Fig. 4-44
Long-necked ground beetle (*Colliuris pensylvanicus*)

Long-necked ground beetle
Colliuris pensylvanicus
(FIG. 4-44)

Biology: The remarkably elongated head and thorax are presumably an adaptation to some intriguing life-history behavior that remains unknown to science. All we can say is that several specimens arrived at our black lights in late April.

Distribution: From coast to coast within the United States, though perhaps spottily so in the far west.

Remarks: Length = 8 mm. This ground beetle is known to favor moist habitats, and the Ottine wetlands would be a good place to fill in the unknown aspects of its biology. Other members of the ground beetle family with elongated forebodies are snail predators adapted to reach within the shell for their prey.

Similar species: None.

Fig. 4-45
Giant tiger beetle
(*Cicindela obsoleta*)

Tiger beetles
Cicindela obsoleta, C. punctulata, Megacephala carolina,
C. belfragei, C. sexguttata
(FIGS. 4-45–4-50)

Biology: All tiger beetles are predators in both the grub and adult stages, and of the four treated here, only Belfrage's species cannot fly. The giant tiger beetle (*C. obsoleta;* length = 24 mm; Fig. 4-45) approaches one inch in length and is the largest of its kind in the United States (Valentine 1947). We found a single specimen near a small ephemeral pond at the interface between a wooded area of seeps and a pasture, and by its large size and dark coloration we thought from a distance that it might be a cricket. This species is wary and tends to fly farther than others when approached. According to one report it flew 420 yards with only three landings across that total distance, and "owing to its size and rapidity it is a formidable foe to other insects" (Leng 1902, 119).

The punctate tiger beetle (*C. punctulata;* length = 14 mm; Fig. 4-46) is much smaller than its relative and much more common. We found adults and vertical larval burrows in late July along the edge of an extinct peat-land that had dried to a barren flat. The burrows were inhabited by grubs that wait at the entrance for passing prey, their heads blocking the entrance and nearly blending in with the surrounding gray soil. When disturbed, they retreat immediately into the depths of the pit. Punctate tiger beetles also occur within city limits and are attracted to lights at night, both habits being rather uncommon among their kind. They are known for occupying relatively hard soils, and at the turn of a recent century

Fig. 4-46
Punctate tiger beetle
(*Cicindela punctu-
lata*)

Fig. 4-47
Carolina tiger beetle
(*Megacephala car-
olina*)

they were still living among the paved streets of downtown New York City (Leng 1902).

We found the metallic green and red Carolina tiger beetle (*Megacephala carolina;* length = 17 mm; Fig. 4-47) hiding during the light of day on the same dry flat where we found the more abundant and diurnally active punctate species. Having otherwise similar ecological niches, the two switch places after dark. Carolina tiger beetles also made appearances beneath the streetlight outside the refectory building at Palmetto State Park.

Fig. 4-48
Belfrage's tiger beetle
(*Cicindela belfragei*)

Fig. 4-49
Belfrage's tiger beetle
in hand after releas-
ing defensive solution

Belfrage's tiger beetle (*C. belfragei;* length = 14 mm; Fig. 4-48) is the most curious and least studied of all. It has wing covers and even microscopic remnants of wings beneath those covers, but of course it cannot fly with such rudimentary equipment. When approached, it runs rapidly into surrounding vegetation. We found them in late spring along trails in Palmetto State Park where they were active during the hottest and brightest hours of the day. We also saw mating pairs at these times. When handled, they bite and release a brown and essentially odorless fluid from the mouth, neither defense having any consequence for humans (Fig. 4-49). They might derive additional protection from their close resemblance to the large black carpenter ants (*Camponotus pennsylvanicus*) sharing this envi-

Fig. 4-50
Six-spotted tiger
beetle (*Cicindela
sexguttata*)

ronment. The similarity is greatest when the beetles scramble away in small groups.

We only rarely found a six-spotted tiger beetle (*C. sexguttata;* length = 14 mm; Fig. 4-50), and only on trails cutting through the floodplain of the San Marcos River. This beetle is said to emit a pleasant fragrance when handled, though we never noticed it.

Distribution: Giant tiger beetles occur from eastern Texas to Arizona with a disjunct population along the Missouri-Arkansas border, punctate tiger beetles and Carolina tiger beetles occur from the Atlantic Ocean to California, and Belfrage's tiger beetle has a more central range suggesting a link with grasslands. It occurs in a narrow north-south band from Nebraska to extreme northern Mexico near Brownsville, Texas. Thus, its range includes most but not all of eastern Texas. The six-spotted tiger occurs from the Atlantic Ocean to a western limit, in the southern part of its range, near the Ottine swamps.

Remarks: We saw only one of these five wetland species in the nearby uplands of the Lost Pines forest (the six-spotted tiger), and this was the only forest species we saw in the wetlands. In the case of the Carolina tiger beetle, the lack of overlap is explained in part by the fact that its grub stage always inhabits burrows located near fresh water (White 1983).

Similar species: The Carolina tiger beetle with its brilliant metallic coloration and typically nocturnal activity is easy to identify, and no other species is as large as the dark giant. However, other tiger beetle species are likely to occur in the area, and formal keys are needed to establish their identity.

Fig. 4-51
Little Texas glow-
worm (*Distremo-
cephalus texanus*)

Little Texas glowworm
Distremocephalus texanus
(FIG. 4-51)

Biology: The biology of the little Texas glowworm is poorly known. In 1881 J. L. LeConte, the greatest nineteenth-century authority on American beetles, reported one of the few observations on this species. By lamplight a male ran across a table, twisting its wings in a peculiar fashion as if trying to straighten them out. When we captured several specimens at our black lights in the Lost Pines forest nearly 120 years later, we saw the same phenomenon. But we did not see any light shining from within the head, nor did we see any at the back end of the little animal as LeConte reported. However, we did see this exactly as LeConte described it while examining a specimen from the Ottine swamps. The grub also shines with a weak light, and it eats snails (apparently in captivity; LeConte 1881).

Distribution: In the United States the little Texas glowworm has been reported only from Texas and Nevada (Zaragoza 1986), though we saw mention of it in a Web page devoted to New Mexico beetles. Specimens in the Texas A&M University Insect Collection suggest that the range of the species in Texas extends from the Panhandle area to the extreme south and then on into Mexico.

Remarks: Length = 6 mm. If the remark in LeConte's publication of 1881 still holds, then the female sex remains unknown. April to early summer is a good time to attract numerous males to black lights.

Similar species: None. It is perhaps notable that the bizarre and much larger glowworms of the genus *Phengodes* were seldom seen here, though dozens were attracted to black lights over the years in the nearby Lost Pines forest.

Fig. 4-52
Haldeman's potato beetle (*Leptinotarsa haldemani*) on its host plant, Texas nightshade (*Solanum triquetrum*)

Haldeman's potato beetle
Leptinotarsa haldemani
(FIG. 4-52)

Biology: Haldeman's potato beetle is a beautiful, dome-shaped, metallic green relative of the less attractive but more famous and pestiferous Colorado potato beetle. It is not a pest itself and is in fact uncommon. It feeds upon leaves of its namesake plant and upon the foliage of related species. We found them on Texas nightshade (*Solanum triquetrum*) as it flowered in early spring.

Distribution: This is one of the relatively few western insects in the wetlands, occurring from Arizona to an eastern limit in east-central Texas (Jacques 1988).

Remarks: Length = 10 mm. We observed mating pairs of Haldeman's potato beetle on their host plants in April.

Similar species: None.

Texas tortoise beetle
Coptocycla texana
(FIG. 4-53)

Biology: The leaf-eating Texas tortoise beetle looks like a small green jewel glowing in the headlamp's beam at night. We found five specimens resting motionless on the undersurface of leaves of rusty blackhaw (*Viburnum rufidulum*). Their only known food plant is anaqua (*Ehretia anacua*; Edward

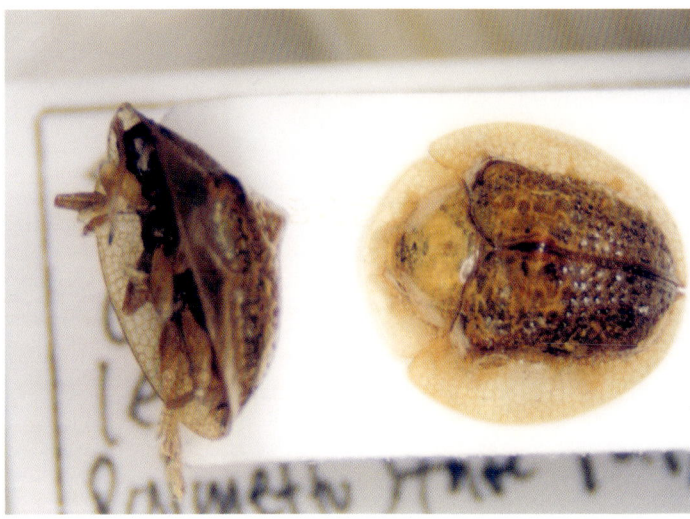

Fig. 4-53
Texas tortoise beetle (*Coptocycla texana*)

Riley, pers. comm., 2003), and because these plants were nearby, the beetles might have been overwintering on blackhaw rather than feeding on its leaves. Or perhaps this is a previously unreported host.

Distribution: Published records indicate that this is a Texas endemic found only in the region from Brownsville (Schaeffer 1933) in the south to Austin in the north (Edward Riley, pers. comm., 2003). However, it probably occurs in Mexico too.

Remarks: Length = 8.5 mm. The breadth of the Texas tortoise beetle's diet and its east-west geographic distribution must still be worked out.

Similar species: None that we saw.

Green dock beetle
Gastrophysa cyanea
(FIG. 4-54)

Biology: This is a small but shiny metallic green beetle visible from some distance as it feeds in groups on foliage of dock and rhubarb. We found them on dock near the oxbow lake of Palmetto State Park. Chemical protection is employed against ants and other enemies in every stage of the life cycle that has been examined. The bright yellow eggs, laid in clusters beneath leaves, are defended by an oily acid that tastes bitter to humans and numbs the tongue. The grub that hatches from the egg is defended by glands on its back that can be everted to dispense a protective chemical. Adults have a more extensive panoply. They feign death, pull in the legs and tumble off the plant, bleed chemicals from self-induced hemorrhage,

Fig. 4-54
Green dock beetle
(*Gastrophysa
cyanea*); mating pair

raise their wing covers and present them to attackers so that the passive
shield becomes actively deployed like that of a human warrior, and secrete
chemicals from pores in this shield (Howard et al. 1982).

Distribution: The green dock beetle occurs from coast to coast within the
United States.

Remarks: Length = 5.5 mm. A related leaf beetle that we did not see in the
swamp builds a case about its body in the grub stage and walks about
inside the shell, resisting attacks from fire ants in a manner not employed
by the green dock beetle.

Similar species: Three close relatives occur in the United States. These are
either not found in Texas or they have a reddish color just in front of the
wing covers.

Red-faced leaf beetle
Anomoea rufifrons mutabilis
(FIG. 4-55)

Biology: Little is known of the red-faced leaf beetle's life history. Its grubs
are remarkable for hatching from eggs enclosed by the mother's frass, for
standing on their heads, and for building a case that envelops and protects their body as they travel about. The first component of the case is
the empty eggshell itself. Over time the structure grows when detritus and
frass are added (Moldenke 1970). Adults might be associated with
shrubby legumes related to mimosa. We found only one specimen, a male,

Fig. 4-55
Red-faced leaf beetle (*Anomoea rufifrons mutabilis*) on switch-grass (*Panicum virgata*) in cordgrass marsh

clinging to switchgrass (*Panicum virgatum*) in the Gulf cordgrass marsh of Palmetto State Park.

Distribution: This short-horned leaf beetle is one of the few southern species we identified in a wetland more eastern in its affinities. The animal occurs from southern Mexico to a northern limit near Amarillo, and except for one Arizona report, within the United States it is known only from Texas.

Remarks: Length = 9.5 mm. It would be interesting to know if this species is biologically associated with switchgrass or if its presence there was only accidental.

Similar species: The subspecies shown here was named for its great variation. Separation of the species from close relatives is difficult.

Oak timberworm beetle
Arrhenodes minutus
(FIG. 4-56)

Biology: The female oak timberworm beetle uses her long beak and the tiny jaws at its tip to dig a deep, tubular hole in the bark of trees. She then deposits an egg in the hole. When the grub hatches, it bores deeper; and when the grubs are present in numbers, the wood becomes riddled with pinholes so that it cannot be used for making barrels and other products. Host plants include oak, poplar, maple, and the species of maple known as boxelder, where we found the beetle in the Ottine swamp.

Males guard egg-laying females by walking around them in circles. They

Fig. 4-56
Oak timberworm
beetle (*Arrhenodes
minutus*)

compete for mates with other males, and battles ensue in which the jaws of the victor might be used to raise the loser overhead and hurl him from the tree. The male's jaws are also used to free the female when her own mandibles become stuck as she chews into bark (Blatchley and Leng 1916).

   Oak timberworm beetles can live as long as four years (Buchanan 1960). Sometimes they form colony-like aggregations of three hundred or more individuals (Sanborne 1983).

Distribution: From the Atlantic Ocean to Texas.

Remarks: Length = 12 mm (but ranges 5–22 mm!). The oak timberworm beetle belongs to a family known as the primitive weevils. Several males arrived at our blacklight traps after dark, though we did not see any reference to this behavior in the literature.

Similar species: None.

Sedge weevil
Listronotus squamiger
(FIG. 4-57)

Biology: We encountered the sedge weevil only in the adult stage when numbers of them flew to incandescent lights after dark. Grubs develop within stems of swamp-dwelling plants (Henderson 1939), including the sedge *Scirpus validus* and various species of arrowhead plants. Their known arrowhead hosts do not occur here as far as we know, and the grubs pre-

Fig. 4-57
Sedge weevil
(*Listronotus squamiger*)

sumably develop within grass-leaf arrowhead (*Sagittaria graminea*) instead. Adults have been found among moss and debris in swamps and at lights after dark.

Distribution: From the Atlantic Ocean to Montana and Texas, though this appears to be the first record for Texas. It is remarkable that *L. squamiger* was not reported from Texas in that species' lengthy, multistate collection record within Henderson's massive 1939 work despite the fact that several Texas endemics were included, some of them new to science, as well as other species collected only a few miles from the Ottine swamps. Mitchell and Pierce (1911) found several related species in nearby Victoria County, but *L. squamiger* was not among them. Nor was it reported from Texas or even the Southwest in general in the checklist of the weevils of North America (O'Brien and Wibmer 1982).

Remarks: Length = 8 mm. This weevil flies and crawls, but we are not sure if it swims in the manner of certain close relatives. These progress just beneath the surface with a kind of rapid dog-paddling stroke (O'Brien 1981).

Similar species: Several very similar weevil species occur in south Texas, and a formal key is required to distinguish one from another.

Fig. 4-58
Ragweed weevil
(*Lixus scrobicollis*)
feeding on its host
plant (*Ambrosia trifida*)

Ragweed weevil
Lixus scrobicollis
(FIG. 4-58)

Biology: This black, gray, and yellow weevil belongs to a genus of stem-boring species that develop as grubs within the stems of herbaceous plants, including the notorious allergenic ragweed (*Ambrosia trifida*). They hollow out stems during their feeding and sometimes form galls just below ground level. We found adults fully exposed and feeding on the leaves of ragweed in late May. Some were coupled as mating pairs. They have also been reported to breed in frostweed (*Verbesina virginica;* Mitchell and Pierce 1911), and we saw adults on these plants as well. Our own observations extend the reported onset of the breeding season from July to as early as May.

Distribution: From the Atlantic Ocean to at least as far west as central Texas.

Remarks: Length = 11 mm. This is an unusually slender though large weevil. A swarm of parasitic insect species is known to attack it (Mitchell and Pierce 1911).

Similar species: The genus *Lixus* contains at least twelve species known to occur in Texas. Weevils in general are notoriously difficult to identify under the best of circumstances. A species easy to recognize for its large size and black and red color is the palmetto weevil (*Rhynchophorus cruentatus*), exceeding 25 mm in length and known from the Atlantic Ocean to Louisiana (Blatchley and Leng 1916; White 1983). We never saw it, nor do we know if it includes dwarf palmetto among its hosts. The grubs feed inside palms and are noisy enough to be detected by sound alone.

Fig. 4-59
Texas checkered beetle (*Chariessa texana*)

Texas checkered beetle
Chariessa texana
(FIG. 4-59)

Biology: The biology of the Texas checkered beetle is unknown, but if the biology of its relative *C. pilosa* holds true, then the grubs live within hickory and other trees and feed upon fellow beetle grubs, such as the redheaded ash borer (White 1983). Adults would also be expected to prey upon a variety of insects.

Distribution: This is a Texas endemic occurring nowhere else. We know of only one previous record, the location where the species was first discovered at the base of the Panhandle in Sweetwater, Texas (Wolcott 1908).

Remarks: Length = 15 mm. Size, shape, and coloration predispose this beetle for misidentification as a firefly. Perhaps there is a genuine mimicry that remains unexplained.

Similar species: Under microscopic illumination dark blue is visible, supposedly distinguishing this checkered beetle from the blacker, more widespread *C. pilosa* (length = 13 mm). Or perhaps there is a single widespread species with previously unrecognized color variation. We found what is clearly the black species atop a brilliantly red and metallic blue relative (*C. vestita;* length = 8.5 mm) in what appeared at first to be a doomed mating effort. But in captivity we found bits and pieces of antennae and legs of both parties, suggesting predation rather than reproduction.

Four-spotted checkered beetle
Pelonides quadripunctatum
(FIG. 4-60)

Biology: The biology of the four-spotted checkered beetle is essentially unknown. The beautiful red and black adults with their antlerlike antennae have been observed on hawthorn flowers (Blatchley 1910) and on the flowers of oak trees. We found several specimens on rough-leaved dogwood flowers in April.

Distribution: From the Atlantic Ocean to Texas.

Remarks: Length = 7 mm. Bright red and black coloration suggests a warning of some kind, but nothing seems to be known of this beetle's chemical defenses, if any do indeed exist.

Similar species: None.

Fig. 4-60
Four-spotted checkered beetle
(*Pelonides quadripunctatum*)

Fireflies
Photinus pyralis, Photuris new species?
(FIGS. 4-61 – 4-62)

Biology: The fireflies we encountered in the Ottine swamps are species that emit light (not all fireflies do). Both sexes are capable of emitting light, though the male's light is typically brighter and more active. This is readily understood when one realizes that the flying male searches out females that often wait near ground level in vegetation. Grubs and even eggs are also luminescent. The function, if any, of light during these nonreproductive stages of the life cycle is unclear. Female fireflies sometimes use their lamps to lure males of other species and eat the latter when they arrive.

Beginning in late April the common firefly (*Photinus pyralis*) can be seen flying at dusk, emitting greenish yellow light from the abdomen. Males have a dipping flight consisting of alternating climbs and drops. The light is lowest at the peaks and brightest in the valleys, thus increasing in intensity as the beetle drops and decreasing as it climbs. Females have a smaller light, fly slower, and often lie in vegetation where they answer the flying male's courtship light with their own. The grub is a luminous "glowworm" that emits light near the end of its lower surface. It crawls along with pushes of its abdomen (McDermott 1910).

As interesting as fireflies are when they light up the spring night with bright green flashes, we were disappointed to discover the great difficulty in identifying them. One female firefly began eating a male of the genus *Photinus* when the two were confined in a container. The female and perhaps even the male might be a new species. The female was flying at midnight on May 5 in a meadow adjacent to the San Marcos River. She was flashing a bright green light. A large male of the same genus but of yet another unidentified species was caught on June 2. The male that the female was preparing to eat might be identifiable but cannot be identified unless he is damaged, because the identification keys require a study of internal structures that must be manipulated.

Distribution: The common firefly occurs from the Atlantic Ocean to New Mexico and perhaps farther west. The distributions of the other fireflies treated here must remain uncertain as long as their identities are unknown.

Remarks: Length = 14 mm. Neither the female nor the male *Photuris* specimen will key out to a recognized species using the standard references (Barber 1951; Green 1956; McDermott 1967). It is not clear which is more likely: a problem with the keys themselves or the presence of one or

Fig. 4-61
Unidentified fireflies; mating pair

Fig. 4-62
Female firefly preying upon male of a different species

more new species in these unstudied wetlands that the keys were never intended to deal with.

Similar species: Additional species of both genera treated here no doubt occur in the wetlands, and, as the previous treatment indicates, identification can be difficult or perhaps impossible.

Fig. 4-63
Bold pleasing fungus beetle (*Megalodacne fasciata*)

Bold pleasing fungus beetle
Megalodacne fasciata
(FIG. 4-63)

Biology: This glossy red and black beetle feeds on fungi in both the adult and larval stages. It has been found on bracket or shelf fungi, under the bark of stumps, at sugar bait, in soil, and at light traps such as the ones that attracted our own specimens. In nearby Victoria, Texas, specimens were taken on a hackberry log and on a Hercules-club tree (Boyle 1956). Adults overwinter beneath bark.

Distribution: Bold pleasing fungus beetles occur from the Atlantic Ocean to Colorado.

Remarks: Length = 15 mm. This attractive beetle is smooth and shiny and could almost be mistaken for a tiny toy.

Similar species: Only one species in the United States bears a resemblance, but the plate behind the head of the giant pleasing fungus beetle (*M. heros*) is more rectangular in outline.

Eyed click beetle
Alaus oculatus
(FIG. 4-64)

Biology: This is the largest click beetle in Texas. When flipped onto its back, it employs a snapping mechanism that throws the animal into the air and back onto its feet. If employed as a defensive escape mechanism while held between the fingers, the force is mildly unpleasant and would surely

Fig. 4-64
Eyed click beetles
(*Alaus oculatus*);
mating pair from
beneath bark of log

cause anyone unacquainted with the animal to drop it in surprise. The large black "eyes" are not eyes at all and are not even located on the head. They make up a second, more passive means of defense that might startle a potential predator. Yet perhaps it goes too far to describe them as examples of "terrifying coloration" (Edwards 1949, 92).

The larval stage of the eyed click beetle is an impressive grub that exceeds 50 mm in length. It lives in decomposing logs where it preys upon other grubs. This carnivorous diet differs from that of the more familiar garden-variety wireworm, also a click beetle, that feeds upon roots instead. We don't know what the adults eat or if they eat at all. They do fly to lights, though they never came to ours, nor did their close relative the blind click beetle (*A. myops*) of the nearby Lost Pines forest. Adults in Indiana are known to appear in numbers in mid-April (Blatchley 1910), and that is precisely when we found our first two specimens in south-central Texas.

Eyed click beetles bite when handled if the opportunity arises. The slender, curved jaws, though sharp, inflict nothing more than a startling pinch.

Distribution: From the Atlantic Ocean to Texas. We never found this species in the Lost Pines where the blind click beetle occurs, and we never found the blind click beetle in the swamp where the eyed species occurs.

Remarks: Length = 45 mm. According to a leading authority, "This is one of the beetles your friends bring to you. It is fairly common and never fails to arouse interest" (Arnett, Downie, and Jaques 1980, 197). One specimen was found essentially underwater, crawling on the artesian well that extends the summer life of the Ottine swamp by supplying a little fluid to at least one of the lagoons in Palmetto State Park. It was plunging indefatigably upward against the strong current of falling water that threatened to carry it away, and it was surprising to find that an insect was able to resist that force. On the other hand, its scientific name does not mean "eyed click beetle" at all. It translates as "eyed wanderer."

Similar species: Only one animal in central Texas is likely to be confused with this one. That species is the blind click beetle (*A. myops*), which occurs in the Lost Pines forest. It bears a greater resemblance to bark and is smaller and browner in color.

5
True Flies and Fleas

Biting flies are not as great a problem in the swamps and marshes of south-central Texas as they are in the northern wetlands of the United States and Canada, where mosquitoes and, especially, black flies are murderous. However, as these words were written, the West Nile virus, transmitted by mosquitoes that feed on birds, was making its first appearance in central Texas.

A second group of biters, the robber or assassin flies, is more apparent here. They do not bite people unless handled and do not take mere blood meals from their prey. Instead, they capture and kill other arthropods and suck their body fluids. Thus, they are full-blown predators, unlike the micropredatory mosquitoes that take blood in the manner of a temporary parasite and do not kill their hosts.

Horse flies
Tabanus atratus, T. gladiator, T. lineola
(FIGS. 5-1–5-3)

Biology: The blue-tailed fly (*T. atratus;* Fig. 5-1) is an enormous blackish fly that attacks livestock. Bloodsucking females probably do as much harm to the animals by annoying them with loud, low-pitched buzzing flights as they do by the biting attack itself. Four to six flies often torment a single cow or horse, but numbers do not increase to plague proportions like those of mosquito swarms (Stone 1930; Schwardt 1932; Teskey 1990). Blue-tailed flies will presumably inflict painful bites on humans too.

We found the female pictured here as she was laying eggs on a dead stem in a cattail marsh. When the maggots hatch, they drop to the wet ground below and prey upon earthworms and other insects, including

Fig. 5-1
Blue-tailed fly
(*Tabanus atratus*);
female laying eggs in
cattail marsh

Fig. 5-2
Gladiator horse fly
(*T. gladiator*)

mosquito larvae and each other, services that should be kept in mind. The pupal stage is often spent inside cells that the maggots themselves construct of mud (Goodwin and Drees 1996).

Distribution: There is disagreement about the extent of the blue-tailed fly's distribution. It probably occurs from the Atlantic Ocean to the Rocky Mountains. The species was previously reported from the Ottine swamp by the authors of the standard reference on Texas horse flies (Goodwin and Drees 1996).

Remarks: Length of *T. atratus* = 24 mm; length of *T. gladiator* = 23 mm;

Fig. 5-3
Lined horse fly
(*T. lineola*)

length of *T. lineola* = 13 mm. The largest species is said to be the blue-tailed fly of American folksong and legend, but unless the light is just right, it is better described as the black-tailed fly. Blue-tailed flies are known to transmit anthrax among livestock (Chvála, Lyneborg, and Moucha 1972).

Similar species: None so large and dark in the Ottine swamp. The brown gladiator horse fly (*T. gladiator;* Fig. 5-2) is a species of slightly smaller size but of much lighter color. Both sexes of this species are notable for appearing after dark at bedsheets hung in association with blacklight traps in the heat of summer. The blue-tailed fly never made such an appearance. Gladiator horse flies occur from the Atlantic Ocean to a new western record in the Ottine swamp, as established here. Their maggots dwell in the upper inch of mud in bodies of still water (Goodwin and Drees 1996). The lined horse fly (*T. lineola;* Fig. 5-3) is smaller than both and lighter in color. We found one female along the banks of a flowing rivulet in a marshy area.

Golden ricefield mosquito
Psorophora ciliata
(FIG. 5-4)

Biology: This huge, beautifully golden mosquito, known to Floridians as the gallinipper (Dozier 1920), prefers open areas and lays desiccation-resistant eggs on moist soil that becomes periodically flooded, including the rice fields of south Texas. When rains come, typically from April through October, larvae (wrigglers) hatch and complete their development in tem-

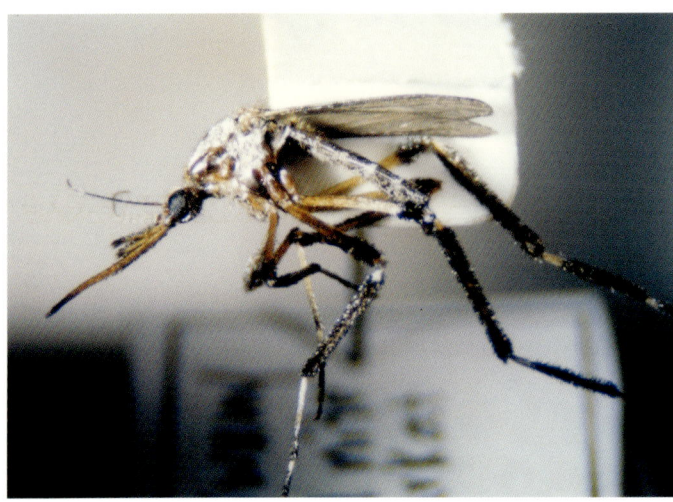

Fig. 5-4
Golden ricefield
mosquito
(*Psorophora ciliata*)

porary standing water, feeding on other mosquito larvae. Mammals are the usual targets of the bloodsucking adult females. Many flew to our black lights in April and early September, and the result is a disturbing sight. With legs trailing behind in flight big females may approach 25 mm in length (Siverly 1972).

Distribution: From the Atlantic Ocean to west Texas with greatest abundance near the coast. We never saw this mosquito during years of study in the Lost Pines forest, where we did see its relative Johnston's mosquito (*P. johnstonii*) and the much smaller inornate mosquito (*Culiseta inornata*).

Remarks: Length = 9 mm. According to standard works on the subject of mosquitoes, "The females are persistent biters, attacking any time during the day when their haunts are invaded and inflicting a painful injury" (Carpenter and LaCasse 1955, 116), and "females are vicious biters" (Bohls 1944, 86). We confirmed these characterizations during a visit to the swamp in April when one of us was bitten on the leg through denim blue jeans. The Texas State Health Department has identified the golden ricefield mosquito as the largest biting species in Texas (Bohls 1944).

Similar species: *Psorophora columbiae,* known as the dark ricefield mosquito, also occurs in south Texas. Mosquitoes are difficult to identify, and formal keys must be consulted.

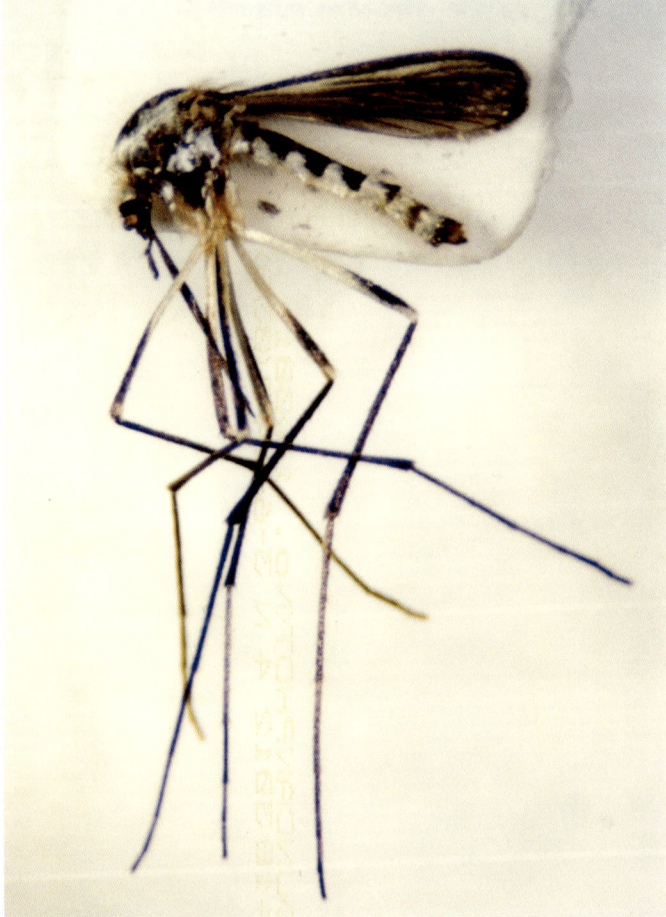

Fig. 5-5
Eastern tree-hole
mosquito (*Aedes tris-*
eriatus)

Eastern tree-hole mosquito
Aedes triseriatus
(FIG. 5-5)

Biology: We captured our first specimen as it attempted to bite at 4 P.M. on
a bright, warm day, and another one week later at about the same time as
it hovered about our heads with half a dozen others in a shaded area. The
eastern tree-hole mosquito has natural enemies in the form of other mosquitoes that prey upon it when all are in the wriggler larval stage. One of
these predators is the huge and beautiful metallic blue *Toxorhynchites*
rutilus. Another is the golden ricefield mosquito.
Distribution: From the Atlantic Ocean to Utah.
Remarks: Length = 5 mm. In Texas at least one close relative of the eastern

tree-hole mosquito has been known to transmit yellow fever (Bohls 1944). We know of no reports linking the disease with the species at hand, but it has been implicated in the transmission of LaCrosse encephalitis, which affects the nervous system and is most common in children.

Similar species: Mosquitoes are difficult to identify, and because identification might be important for medical reasons, an authoritative text should be consulted. The well-illustrated work of Carpenter and LaCasse (1955) and the more dated and less exhaustive compilation of Texas species (Bohls 1944) are good places to start.

Fig. 5-6
Phantom crane fly (*Bittacomorpha clavipes*) hanging from fern leaf above the muck of Rutledge Swamp

Phantom crane fly
Bittacomorpha clavipes
(FIG. 5-6)

Biology: Phantom crane flies strike the eye as tiny specters when they are first seen flying slowly in an attitude that might be described as "standing up," with all six of their black-and-white striped, hair-thin legs spread out full-length all around, like the spokes of a wheel (Alexander 1981, 325). The sight might even conjure up the image of a flying spider's web, or "spots before the eyes" (Cole 1969, 72). According to one observer, "This species is one of the most conspicuous and interesting of all Nearctic Diptera. The first tarsomere of the legs of this species is dilated and filled with tracheae, a characteristic which enables the flies to drift in the wind with their long legs extended to catch the breeze" (Alexander 1981, 326).

Like other crane flies these do not bite, though they might be mistaken for huge mosquitoes. From time to time they land on vegetation to grasp a leaf or stem with the front legs and hang there as if attempting a pull-up on an exercise bar. Two or three may be observed bumping into one another at these sites, perhaps competing for a favored spot. An earlier student of phantom crane flies reported that the maggot stage can be found "in the early spring in a shallow swampy slough full of rushes and swamp grass" (Howard 1904, 95–96). This mirrors our own experience, for we saw the adults only in such habitats, first in a cattail marsh and later in Rutledge Swamp among water hemlock and windthrows.

Female flies deposit hundreds of eggs in flight in clutches of three to five by dipping their abdomen in water (Bowles 1998). The maggots live in muck itself and are tubular and rusty brown in color, resembling bits of decaying grass stem. This might confer protection by camouflage as they feed upon diatoms and plant tissue while breathing through a long tube or siphon.

Distribution: This particular phantom crane fly species occurs from the Atlantic Ocean to Arizona.

Remarks: Length = 15 mm. If observers do not have a mental search image in place, they could easily overlook the phantom crane fly because of its slender form; its preferred flight paths through thick, lush vegetation; and its tendency to stay fairly close to the surface of swampy ground. We never noticed them until one was sighted near dusk, with the setting sun *behind* it, thus providing a serendipitous backlight that accentuated its unusual form.

More than a century ago an observer in Illinois summarized his experience and ours by noting, "Their singular and ghostly appearance as they floated slowly by with their black and white legs radially extended will never be forgotten" (Hart 1895, 192).

Similar species: None in the swamps of south-central Texas.

Common crane fly
Genus and species undetermined
(FIG. 5-7)

Biology: On March 15, we saw thousands of large crane flies in the broad daylight of afternoon in Palmetto State Park, some mating on boxelder leaves and others resting or mating on tree trunks. This was the largest bloom of crane flies we witnessed. They seemed to be everywhere in the swamp.

Distribution: Uncertain without species identification.

Fig. 5-7
Crane fly, undeter-
mined species killed
by fungus on dwarf
palmetto

Remarks: Length = 27 mm. Crane fly larvae are known as leather jackets, and many of them feed on the roots of plants. The genus shown here might be *Tipula*.

Similar species: There are many crane fly species in Texas, and as of this writing there was no key allowing identification to genus or species level, indicating a remarkable lack of interest in these abundant animals that are so large and so closely resemble mosquitoes. We thank George Byers for alerting us to this fact.

Black crane fly
Gnophomyia tristissima
(FIG. 5-8)

Biology: Adults are known to frequent wildflowers, and that is where we captured our own specimen, which has a large mite attached to its body. It isn't clear if this arachnid is a parasite or if it was merely catching a ride. The fly's maggot stage lives and feeds within nearly liquefied rotten wood of decomposing logs (Rogers 1928). This is also the resting site for the pupal stage between larva and adult. Little else is known about this small and very dark species.

Distribution: From the Atlantic Ocean to central Texas.

Remarks: Length = 6 mm. It is said that this fly species is more likely to be seen in herbaceous vegetation than its close relative discussed in the next section.

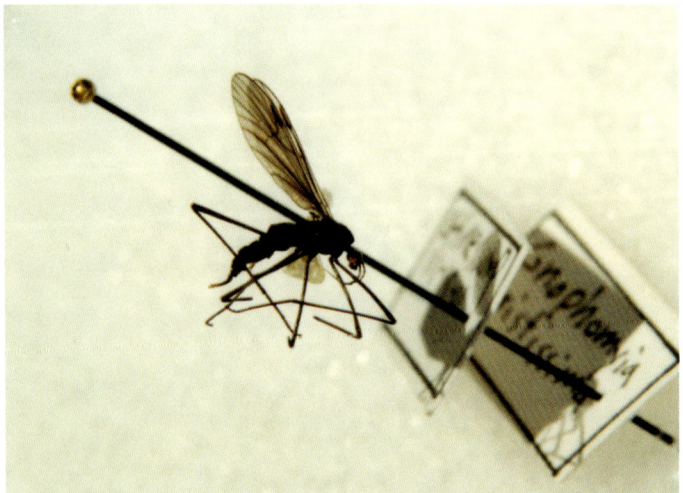

Fig. 5-8
Black crane fly
(*Gnophomyia tristis-sima*)

Similar species: A very close relative, also black and much smaller than the huge common crane flies, might occur in the swamp. This species is *G. luctuosa*, which differs by its possession of a patch of tiny hairs near the apex of the wing. Though it can be distinguished only with the aid of a hand lens or microscope, this fly has a different ecological niche, breeding in decomposing parts of living, standing trees rather than in similar micro-habitats of dead, fallen trees (Rogers 1928).

Striped stilt fly
Taeniaptera trivittata
(FIG. 5-9)

Biology: The biology of the striped stilt fly is unknown. Adults resemble parasitic wasps and were previously reported to fly in May and June (Arnett 2000), but in the Ottine swamp we found both sexes flying as early as April in two consecutive years. Stilt flies in general are believed to develop as maggots in dung or in decaying plants, and the adults of at least some species visit and presumably feed upon excrement in moist areas. Some stilt flies, including the one featured here, perch on foliage, such as that of frostweed, and wave the white-tipped front legs in a dancelike display that suggests the waving of semaphore flags on a landing strip.

Distribution: From the Atlantic Ocean to a western limit in central Texas.

Remarks: Length = 10 mm. When we first sighted these flies, we mistook them for the parasitic wasps that stilt flies are said to mimic in appear-

Fig. 5-9
Striped stilt flies
(*Taeniaptera trivit-tata*); mating pair

ance. The female's resemblance to its model is greater than the male's.

Similar species: None so far north, but several close relatives occur in the Brownsville area of extreme southern Texas.

Giant robber fly
Microstylum morosum
(FIG. 5-10)

Biology: Giant robber flies perch on twigs and swoop down on grasshoppers as large as themselves when advancing boots flush the hoppers out. We watched one female hunting in a marshy area, and her behavior was like that of a hawk that flies in pursuit of sparrows. However, out of more than half a dozen attempts she failed to capture a single specimen of the big differential grasshopper (*Melanoplus differentialis*). Cicadas are also taken as prey. Our experience confirms an earlier description: "Its huge size, dark color, and emerald-green eyes impart an antedeluvian [*sic*] appearance as it rests on a dead branch of a tree, its intent watching for prey betrayed by the quick pivotal movements of the head surmounting the stiltlike neck" (Bromley 1934, 97). On the last day of June we saw several flying about, including one mating pair, in the Palmetto Trail area where we had not noticed the species before.

Distribution: This "magnificent animal," once considered a Texas endemic species, is now known to occur from an eastern limit in Texas not far from the swamp to a western limit in Arizona (Beckemeyer and Charlton 2000).

Fig. 5-10
Giant robber fly
(*Microstylum moro-sum*)

Remarks: Length = 50 mm. The giant robber fly is perhaps the largest of its kind in North America and is only a few millimeters smaller than the largest fly in the world. It is also known as "the green-eyed monster" (Bromley 1934, 97) and is easier to capture than most. This fatal arrogance might be an ironic consequence of its own size and power.
Similar species: None.

Dark bee-killer
Saropogon dispar
(FIG. 5-11)

Biology: This large, dark robber fly is a pest of beehives in south-central Texas, where it flies erratically with a dull buzzing sound as it seeks unmated queens as favored prey. If worker bees detect its approach, they mob the invader and sting it to death, leaving evidence of the fight in the form of dead flies lying on the ground near the hive. Other prey include grasshoppers, mud daubers, and red paper wasps as large as the fly itself (Bromley 1934). One of our specimens was feeding upon a male wasp of a species that forms sleeping aggregations on wax myrtle.
Distribution: Oklahoma and Texas.
Remarks: Length = 29 mm. A dark bee-killer was once observed while attempting to kill and eat a boll weevil. After ten minutes and numerous failed efforts to pierce the hard shell with its beak, it abandoned the cotton pest, leaving it alive but covered with saliva (Bromley 1934). Such encounters go a long way toward explaining the evolutionary success of

Fig. 5-11
Dark bee-killer
(*Saropogon dispar*)
with unidentified
prey underneath

beetles, even when compared to other insects. The honeybee that it favors is not native to the United States and thus provides an example of an exotic animal having favorable effects upon a native species. A beekeeper in the San Antonio area recommended wooden paddles to keep the predators under control by swatting them as if they were giant house flies.
Similar species: None that we saw.

Bumblebee mimics
Mallophora orcina, Laphria macquartii, L. flavicollis
(FIGS. 5-12–5-14)

Biology: These three hairy robber flies have yellow and black coloration and, in the case of the first two species, a size and shape that combine to make them easily mistaken for bumblebees. The similarity might be an example of "aggressive mimicry" that allows predators to more readily approach the prey that they resemble. *Mallophora orcina* does indeed capture and kill its powerful model (Fattig 1933) and feeds almost exclusively on these and other stinging insects. Little *L. flavicollis* favors beetle prey, but the prey of *L. macquartii* is unknown.

Distribution: *Mallophora orcina* probably occurs at least sporadically from the Atlantic Ocean to central Texas, *L. macquartii* is a Texas endemic, and *L. flavicollis* is an eastern species that occurs at least as far west as central Texas. There has been speculation that Texas *M. orcina* specimens represent a previously unrecognized species.

Remarks: Length of *M. orcina* = 24 mm; length of *L. macquartii* = 27 mm;

Fig. 5-12
Bumblebee mimic robber fly (*Mallophora orcina*)

Fig. 5-13.
Bumblebee mimic robber fly (*Laphria macquartii*)

Fig. 5-14
Bumblebee mimic robber fly (*Laphria flavicollis*) with scarab beetle prey

length of *L. flavicollis* = 19 mm. *Mallophora orcina* has bright yellow "fur," whereas the other two bee mimics have dingier coloration. *Laphria flavicollis* is much smaller than all bumblebees familiar to us and has a black abdomen with little or no yellow.

Similar species: None.

Fig. 5-15
Gloomy robber fly
(*Orthogonis stygia*)

Gloomy robber fly
Orthogonis stygia
(FIG. 5-15)

Biology: The biology of the gloomy robber fly is unknown beyond its seasonal appearance as an adult in the summer, as both Bromley (1931) and we have reported (Taber and Fleenor 2003a). We found a single specimen, a female, on a trail in Palmetto State Park. The male sex remains unknown to science. At 27 mm in length and with a black body and black and violet wings, this predatory fly profits from a strong resemblance to spider wasps that defend themselves with a powerful sting.

Distribution: This appears to be the first published record of the species from Texas, and the previously known distribution was highly disjunct from the Ottine swamp, for it was limited to North Carolina, Mississippi, and Florida. The Texas A&M University Insect Collection contains three specimens collected in June and July in Liberty County, Texas, well east of our wetland site.

Remarks: Length = 27 mm. *Stygia* refers to this predator's gloomy aspect and is a term associated with the River Styx of Hades. This is the only species of the genus in the United States. Its discoverer, an expert on robber flies, wrote, "The species appears to be confined to Southeastern United States and is undoubtedly very rare" (Bromley 1931, 434).
Similar species: None.

Fig. 5-16
Bicolored robber fly
(*Lampria bicolor*)

Bicolored robber fly
Lampria bicolor
(FIG. 5-16)

Biology: Elsewhere, this attractive predator frequents logs and stumps of post oaks (Bromley 1934). We found our single specimen on the leaf of a bumelia tree where a variety of interesting insects were congregating.
Distribution: Occurs in the eastern United States with uncertain limits.
Remarks: Length = 18 mm. The front half of this fly is black, including the wings, and the abdomen is a strongly contrasting red.
Similar species: None.

Fig. 5-17
Orange bee-killer
(*Diogmites symmachus*) perched on ironwood twig (*Sideroxylon lanuginosum*)

Orange bee-killer
Diogmites symmachus
(FIG. 5-17)

Biology: Orange bee-killers are big, orange-brown, slow-flying predators with bright green eyes and long legs that dangle beneath the tubular body like grappling hooks as they cruise through meadows in search of food. They pounce on other insects, including fellow robber flies and that other "bee-eater," the assassin bug *Apiomerus crassipes*. When tackling dangerous prey such as big red paper wasps, the bee-killer uses its six gangly legs to hold the probing stinger at a distance from its body until exhaustion overcomes the prey. Then the beak delivers the coup de grâce with a stab in the eye or in the soft membrane between hard exoskeletal plates. The bee-killer injects a toxin, sucks out the body fluids, and discards the dry husk. We found the pictured specimen on a perch amid the foliage of a bumelia tree.

Distribution: The bee-killer has an unusual range within a narrow corridor of the United States, including only Texas and Louisiana in the south, but extending north across the plains to Canada.

Remarks: Length = 32 mm. Long ago an authority on this animal and its relatives declared Texas to be the state with the greatest diversity of robber flies in the entire country (Bromley 1934).

Similar species: None that we saw.

Fig. 5-18
Three-banded rob-
ber fly (*Stichopogon
trifasciatus*)

Three-banded robber fly
Stichopogon trifasciatus
(FIG. 5-18)

Biology: This tiny species is a predator in the adult stage and presumably in
the larval stage as well, though the latter's diet is not known. Adults prefer to feed upon small jumping spiders (Bromley 1934). We saw only a
few individuals, all of them in the Soefje wetlands and always on dry, bare
spots surrounded by trees, shrubs, or pasture grass.

Distribution: Across much of the United States (Wilcox 1936).

Remarks: Length = 15 mm. Three-banded robber flies are small and uncommon, but they are easy to find because the bare spots they prefer are conspicuous. Inclusion of spiders in the adult's diet is remarkable because
most "murder flies" seem to take insects, winged ones in particular, and
often while the prey is in flight.

Similar species: None that we saw.

Hine's robber fly
Promachus hinei
(FIG. 5-19)

Biology: This species occurs in river bottoms and preys upon red paper
wasps, honeybees, and even the orange bee-killer (see previous discussion),
which would seem a match for it (Bromley 1934). Its success in this
instance might require a literal adherence to the origin of its name, for
promachus means "first to attack."

Fig. 5-19
Hine's robber fly
(*Promachus hinei*)

Distribution: From Ohio to Texas.

Remarks: Length = 36 mm. Hine's robber fly has been described as active, wary, and prone to fly with a shrill, buzzing sound (Bromley 1934).

Similar species: No other species that we saw has a series of alternating black and gray bands extending the length of the abdomen.

Round-faced robber fly
Scleropogon subulatus
(FIG. 5-20)

Biology: This species is said to frequent grasslands and open woods in sandy uplands (Bromley 1934), though we saw a single specimen in the swamp. If what is known about related flies is true of this one as well, then the diet is likely to include grasshoppers, crane flies, and its own kind (Wilcox 1971).

Distribution: From the Atlantic Ocean to Texas.

Remarks: Length = 27 mm. This slender robber fly is common in sandy Texas uplands and is probably more abundant in the nearby Lost Pines forest.

Similar species: None.

Fig. 5-20
Round-faced robber-fly *(Scleropogon subulatus)*

Fig. 5-21
Tiger bee fly (*Xenox tigrinus*)

Tiger bee fly
Xenox tigrinus
(FIG. 5-21)

Biology: The tiger bee fly is probably a "parasitoid" because it is believed to be parasitic in one stage of its life cycle (the larva or maggot) but nonparasitic in another stage (the winged adult). If the life cycle of perhaps its closest relative is an accurate guide, maggots develop on stored pollen within the nests of carpenter bees (genus *Xylocopa*) until the bee grub hatches from its egg. At this time the juvenile bee fly attaches to the grub and begins feeding parasitically upon it (Hurd 1959). If the host should manage to shrug off its enemy, the parasite will search until it finds the host and reattaches itself. Eventually the maggot kills the grub and transforms into an active pupal stage that wriggles through the bee's nest with the aid of hooked hairs until it finds the exit. Here it leaves the pupal skin behind and flies away to find a mate and continue the life cycle of the species. Its huge wings (45 mm from tip to tip) display an attractive pattern that has been compared to Spanish lace (Cole 1969).

Distribution: From the Atlantic Ocean to an undetermined western limit somewhere in Texas.

Remarks: Length = 18 mm. In Michigan this large, attractive fly is sometimes tame enough to be scooped up by hand from among the cattails with no need for a net. In fact, "these flies display much curiosity concerning people, often hovering about the neck or arms of people and alighting on them" (Hull 1973, 440). Our experience in the Ottine swamps confirms this behavior.

Similar species: There is much confusion about the scientific names appropriate for this bee fly and its relatives (Hurd 1959; Marston 1970; Hull 1973; Evenhuis and Greathead 1999). However, based upon currently understood geographic distributions, of the five species currently recognized in the genus *Xenox,* only the tiger bee fly is expected to occur in central Texas.

Mydas fly
Mydas clavatus
(FIG. 5-22)

Biology: Very little is known about the biology of the large, rare, and rather startling mydas fly. Its unsettling appearance is attributable to its size and shape and especially to its mimicry of spider wasps. We saw only four mydas flies in the swamp, all in late May and early June and all on or just off the Palmetto Trail in Palmetto State Park. One settled briefly on a log at the edge of a dry lagoon before flying away, two others were captured, and a fourth, probably the mate of one of the latter individuals, escaped our nets. Their resemblance to robber flies has earned the adults a reputation as predators, but this is not the case. They feed on flower nectar instead (Zaitlin and Larsen 1984; Williams 1995).

Males of a related species, and perhaps this one too, stake out mating territories, sometimes at the tops of hills, and defend them against other males (Cooper 1981; Nelson 1986). On these occasions the negative side of a protective similarity to wasps is revealed because the latter also use

Fig. 5-22
Mydas fly (*Mydas clavatus*)

the hilltops, and much energy is needlessly spent by male flies when they challenge male wasps mistaken for competing male flies. Of course, the wasps are looking for their own females as well.

Mated female mydas flies lay their eggs where partially buried wood is exposed at the surface of sandy soil (Gibson 1965). Maggots hatching from these eggs probably live in the soil or in the wood itself, where they prey upon beetle grubs.

Distribution: From the Atlantic Ocean to at least as far west as central Texas, though the distribution, like the biology, remains uncertain.

Remarks: Length = 33 mm. The paucity of information on such a large, attractive fly is difficult to understand, but perhaps it is due to the fly's relative scarcity.

Similar species: The size and shape of the big, dark insect predispose it for confusion with the largest of the robber flies. An orange band on the dark abdomen serves to distinguish it from all such species. The spider wasps that they resemble have four wings rather than two.

Black and gold flower fly
Meromacrus acutus
(FIG. 5-23)

Biology: We captured a single specimen of this beautiful species flying near rough-leaved dogwood flowers at the edge of Rutledge Swamp. The biology of the adult and, especially, the larval stage appears to be unknown.

Fig. 5-23
Black and gold flower fly (*Meromacrus acutus*)

Fig. 5-24
Bumblebee flower fly (*Teuchocnemis bacuntius*)

Distribution: From the Atlantic Ocean to Texas and south into the Tropics.

Remarks: Length = 18 mm. Our encounter was in late April. In Austin, Texas, we found the species at nearly the same time of the month though several years earlier, feeding at flowers of Texas colubrina shrubs (*Colubrina texensis*).

Similar species: None.

Bumblebee flower fly
Teuchocnemis bacuntius
(FIG. 5-24)

Biology: We caught several specimens near Rutledge Creek in April. They nectar at rough-leaved dogwood (*Cornus drummondii*) and rusty blackhaw (*Viburnum rufidulum*), and their buzzing flight is so rapid that capture is difficult. The biology of adult and larva seems to be largely unknown.
Distribution: From the Atlantic Ocean to Texas.
Remarks: Length = 17 mm. Males have remarkably swollen hind legs with a roosterlike spur that might be used for fighting other males or for mating with females.
Similar species: A few robber fly species also resemble bumblebees, but their mouthparts are shaped for piercing rather than for sponging pollen and nectar.

Deer ked-fly
Lipoptena mazamae
(FIG. 5-25)

Biology: In North America the adult stage of the deer ked-fly is a tiny, external bloodsucking parasite of white-tailed deer (*Odocoileus virginianus*). In South America it attacks a deer species known as the brocket (*Mazama americana*), and on a few occasions it has been taken from cattle, collared peccaries (*Tayassu tajacu*), and the weasel-like tayra (*Eira barbara*), although these last three are not considered to be normal hosts (Maa 1969).

Fig. 5-25
Deer ked-fly
(*Lipoptena mazamae*)

Adult flies have wings until they settle on a host, at which time they shed at least a portion of the flight apparatus and are no longer able to fly. An interesting adaptation to parasitic life is the manner in which the young develop. The maggot hatches from the egg while it is still inside the mother's body; and even this stage delays its exit, developing within the mother's uterus, where it feeds on secretions described as a kind of milk, until the time comes to transform into the transitional stage between larva and adult. At this point the maggot drops from the host's body and matures within the soil via the pupal stage (Maa and Peterson 1987).

Distribution: This is a widely distributed parasite, ranging from southern South America to a northern limit in the southeastern United States.

Remarks: Length = 4.5 mm. The deer ked-fly has been previously reported from Gonzales County within a few miles of the Ottine wetlands (Bequaert 1957). Our own encounter was serendipitous, because the parasite somehow found its way to one of our arms. If it had bitten, it would have been the first such record of an attempt to use a human as a host.

Similar species: None, although a microscope is required to identify these tiny animals and their relatives.

Soldier flies
Stratiomys undetermined species
(FIG. 5-26)

Biology: We encountered the larval or maggot stage in an ephemeral pond along the north fork of Rutledge Creek. It was summer, and the shallow

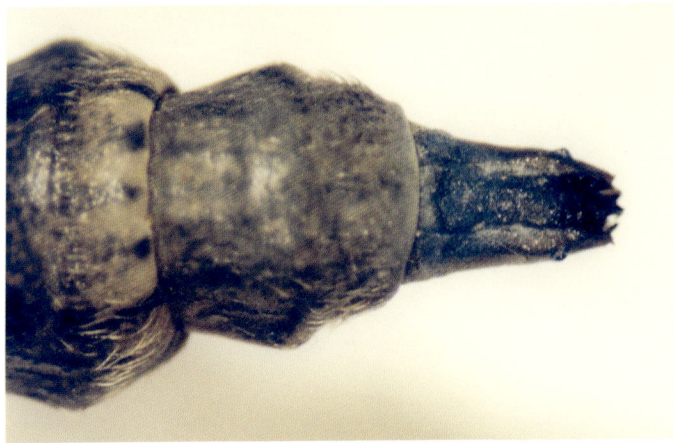

Fig. 5-26
Soldier fly maggot (undetermined species); anterior end

water was so hot that we were surprised to find maggots surviving at such high temperatures. They were well over 25 mm in length, and each was hanging upside down in the water column with the breathing pore, located at the tip of the long "tail," in contact with the surface. When disturbed, they swam down into the mud. Though their appearance and behavior suggest a predatory life, their food consists of small particles instead.

When the time arrives to transform into the beelike adult, the intermediate pupal stage decreases dramatically in size while remaining within the old larval skin, wrapped within a silken cocoon. As a result a large air space forms inside the larval shell, and the whole assembly floats on the surface of the pond until a breeze pushes the biological boat ashore (McFadden 1967). On land the adult may safely emerge and begin flying and feeding upon nectar. Female flies complete the cycle when they lay their eggs on plants at the water's edge. A likely place to find them is on grass-leaf arrowhead (*Sagittaria graminea*).

Distribution: This soldier fly genus occurs widely in the United States, but species cannot be reliably determined without the winged adult stage.

Remarks: Length of larvae = 38 mm. In Colorado a population of soldier fly maggots was discovered in a hot spring at a temperature of 157°F (Packard 1882). The common name is derived from the attractive color patterns displayed by the more familiar adult stage.

Similar species: Aquatic habitat, long body, and tapering siphon are characters found in some other larval flies. Formal keys should be consulted to identify these forms to the genus level, but there are no keys that allow identification of individual species. The maggots at hand might be those of *Stratiomys meigenii.*

Swamp rabbit flea
Euhoplopsyllus affinis
(FIG. 5-27)

Biology: We encountered the swamp rabbit flea in spring in Palmetto State Park, where it leaped onto one of our arms as we explored one of the hiking trails. Its natural hosts include jackrabbits, cottontails, foxes, bobcats, raccoons, rats, mice, and cats (Hubbard 1947; Eads 1950).

Distribution: This is one of the relatively few western animals of the swamp, occurring from Arizona, where it was first discovered, to an eastern limit in Iowa. Though it was previously reported from Texas, our source did not include a report from the Ottine region (Eads 1950).

Fig. 5-27
Swamp rabbit flea
(*Euhoplopsyllus affinis*)

Remarks: Length = 2.5 mm. We know of no records of this flea biting humans, nor did it bite us.

Similar species: None, although the identification of fleas requires a microscope and formal keys.

6
Grasshoppers and Crickets

The lush vegetation of the Ottine wetlands is an ideal habitat for attractive green katydids with remarkably long and filamentous antennae. For this reason katydids are also known as long-horned grasshoppers just as certain beetles that dwell inside the Ottine trees rather than upon their trunks and branches are known as long-horned wood borers. Katydids are not nearly as abundant in the post oak savannas and agricultural lands beyond the protective isolation of these relict swamps and marshes. Remarkably, some species are not gentle herbivores that feed upon leaves. They are voracious predators that attack, kill, and eat other grasshoppers.

Round-headed katydids
Amblycorypha oblongifolia, A. uhleri
(FIGS. 6-1–6-2)

Biology: Swamp katydid adults (*A. oblongifolia*) appear in large numbers in mid-May on lush, leafy, herbaceous vegetation. Their food plants include goldenrod and unidentified grasses (Blatchley 1920), honeysuckle, tobacco, huckleberry (Rehn and Hebard 1914), and various unidentified vines, bushes, shrubs, and "weeds." We even saw them eating poison ivy at night. In the Ottine swamp, Florida lettuce (*Lactuca floridana*), reaching heights of ten feet or more, is a favorite forage sometimes defoliated by herbivorous herds of swamp katydids, which strip the leaves like plague locusts in a field. Our own experience confirms the generalizations of other observers who recorded the katydids' affinity for low moist ground near sources of water. Adults can be captured by hand, but if the approach is not a stealthy one, they will escape with a short flight to a neighboring plant.

Fig. 6-1
Swamp katydid
(*Amblycorypha
oblongifolia*); mating
pair on leaf, with
female behind male,
feeding on glandular
secretions

Fig. 6-2
Uhler's virtuoso katy-
did (*Amblycorypha
uhleri*)

The male swamp katydid courts females by rubbing the front wings together to produce a song. One listener compared the sound to a comb being drawn across a taut string (Blatchley 1920). Others described it as "kizizik," or, more imaginatively, as "katydid." Once mated, the females climb down from their perch in the vegetation and lay their eggs in moist soil.

Uhler's virtuoso katydid (*A. uhleri*) is one of the smallest members of its genus, with tiny wings measuring hardly more than 25 mm in length. We found it to be rare as well. A single specimen, a male, was collected in the

Ottine wetlands, whereas none were seen in the nearby uplands of the Lost Pines forest.

Very little of this animal's biology is known because at the time of writing, what had long been considered to be a single widely distributed species was being split into several, and most of what was known about these far-flung populations had been observed too far east to represent what is now known as Uhler's virtuoso katydid of central Texas. However, an early report of the species from vines along a river bottom in Kerrville, Texas, still applies today (Rehn and Hebard 1914).

Distribution: The swamp katydid occurs from the Atlantic Ocean to Colorado, whereas Uhler's katydid has a very small range indeed. In fact, it is a Texas endemic restricted to the very heart of the state, from Dallas in the north to Uvalde in the west and south, to the Bryan–College Station area in the east.

Remarks: Length of *A. oblongifolia* = 25 mm; length of *A. uhleri* = 17 mm. We saw more katydids in the Ottine swamps than we saw anywhere else in central Texas. At peak abundance hundreds of swamp katydids can be sighted in a few minutes as an observer walks through dense vegetation. It should be noted in passing that both species treated here are among the many "false" katydids that are encountered much more commonly than the few true species that prefer trees to herbaceous plants.

In the eastern United States swamp katydids exist in a beautiful pink form that we never saw. Presumably it occurs here too. On the other hand, yellow individuals of both sexes are common in this part of the state. An earlier report on the species mentioned green, pink, brown, tan, and yellow specimens but noted that only one yellow individual had been reported until that time (Hancock 1916).

More information about the relatively rare Uhler's katydid is needed. Its small size and restricted Texas distribution should greatly aid in identification.

Similar species: One close relative of the swamp katydid bearing a similar appearance occurs in the drier uplands of central Texas. That species is the Texas false katydid (*A. huasteca*). It is best distinguished by the use of formal keys (Rehn and Hebard 1914; Blatchley 1920). We follow Blatchley's concept of the swamp katydid, which combines two of Rehn and Hebard's species under one name. On the other hand, no other member of the genus known to us in the area is as small as the Uhler's species. The true katydid (see following discussion) is easily distinguished from all three false katydids by its wings, which are not flattened against the side of the body but seem to surround the abdomen like a rounded shell.

Fig. 6-3
True katydid (*Ptero-phylla camellifolia*)

True katydid
Pterophylla camellifolia
(FIG. 6-3)

Biology: The true katydid is a big, green, attractive grasshopper with long, threadlike antennae and forewings that look incredibly like the leaves that it eats and clambers among. These "organs of flight" are used for camouflage, gliding, and singing, but paradoxically they are never used for sustained flight. Even clear membranous hind wings are used only as an aid in gliding.

True katydids prefer the heights of oaks and other trees and are thus heard more often than seen. Males do the singing in the strict sense of the word, though females can produce sounds by rubbing their wings together and do so when disturbed or when responding to males. After dark in June the woods in and around the Ottine swamp resound with the "gruck-gruck-gruck" of calling males, a rather harsh sound that can also be described as a shuddering chatter. One student of the subject enjoyed the song as much as we do: "I have often heard the night air simply ringing with their song in the lofty river forests of the southeast, all of the singers being virtually inaccessible in the tree-tops, well over 100 feet above" (Hebard 1941, 198). These arboreal grasshoppers have more than one tune in their repertoire, and on occasion the listener might hear something vaguely resembling the "katy-she-did" that gave katydids their name. Once mated, the females lay their eggs on or in bark. We saw one female doing this near midnight on the trunk of an ash tree.

True katydids are abundant in the field but rare in collections because of their preference for the heights of trees and the difficulty of collecting them there (Hebard 1941). We were lucky to encounter them from time to time on low-growing herbaceous plants such as frostweed and lettuce. Perhaps these were individuals that fell from oak, ash, and pecan branches high above. On several occasions we sighted singing males in boughs not too far from the ground and adopted Hebard's method of collecting them. We tapped the branch whereon it perched with a long stick that was lying nearby, and the insect glided down to the ground just as Hebard predicted it would.

Distribution: The true katydid occurs from the Atlantic Ocean to a western limit very near the Ottine swamp.

Remarks: Length = 50 mm. The Ottine swamp is a particularly fine place to see the true katydid because in June it is remarkably common on low-growing herbaceous vegetation, especially the females, which are said to be less commonly encountered than males. The literature is nearly unanimous with the contrary view that elsewhere these animals tend to stay high in the trees beyond the range of human vision.

Similar species: No other species that we saw in the wetlands has convex wings that cover the abdomen like the valves of a clam. This feature is one of the marks of a true katydid.

Long-spurred meadow katydid
Orchelimum silvaticum
(FIG. 6-4)

Biology: These false katydids are common in summer in the flooded ash swamp of Soefje Swamp, where they perch, sing, and fly from emergent herbaceous vegetation. Others have described the long-spurred meadow katydid as clumsy and easy to capture. We found males and females clinging to long stems and can agree with the latter characterization. In the eastern United States they are found high in post oaks as well as on corn and in lush grass, but we have not seen them in trees. Meadow katydids in general seem to be omnivorous, eating flowers, foliage, and other insects, such as moths, beetles, and sometimes one another (Blatchley 1920). The song is a soft chatter that is hard to describe. When the species was first made known to science, the call was described as a series of "zips" and "zees." Mated females chew holes in the stems of herbaceous vegetation and lay their eggs inside.

Distribution: From Pennsylvania to the Pecos River area of west Texas.

Remarks: Length = 23 mm. In life these attractive grasshoppers have striking orange-red eyes.

Fig. 6-4
Long-spurred
meadow katydid
(*Orchelimum sil-
vaticum*)

Similar species: Several meadow katydids occur in central Texas, and the
best way to distinguish one species from another is by a microscopic
examination of the male's abdomen.

Greater arid-land katydid
Neobarrettia spinosa
(FIGS. 6-5 – 6-6)

Biology: The greater arid-land katydid is a big, green, attractive, rarely seen
predator that kills and eats other grasshoppers. It has a defensive behavior
or display much like that of another predator, the mantis; when disturbed,
it raises its gangly, spiny front legs into the air and waves them about. The
remarkably reduced hind wings, useless for flight, are colored in a light
and dark checkerboard pattern that becomes part of the threat when
expanded. This katydid will box the intruder and bite with powerful jaws
if the annoyance should escalate to contact.

   The preferred habitat of the greater arid-land katydid is mesquite
savanna, where it dwells in trees, bushes, and shrubs. Males can be heard
at night singing from these perches, which is how we first located them on
branches and trunks of mesquite near the Ottine swamp but in the more

Fig. 6-5
Greater arid-land
katydid (*Neobarret-
tia spinosa*); female

upland habitat favored by the spiny tree. The call is a series of rasps
about one second apart, accompanied by an upward jerking of the body as
the wings spread apart to make the sound. We discovered that captive
specimens sing during the day. This usually consists of rasps separated by
a few to many minutes, but sometimes ten or eleven follow one another
with only a second or less between. They will call even when the observer
draws near and stares closely.

Surprisingly little has been written about this predator's feeding habits.
In the field they have been seen eating other grasshoppers and a cicada
(Cohn 1965). Dietary tests demonstrated that flesh is preferred but that
water-rich plant food is sometimes accepted (Gangwere 1990). Even small
amphibians, such as frogs and toads, were apparently eaten in food preference tests. Here are some notes we made while watching one particularly
hungry male captured near the Ottine swamp: We offered a dead false
katydid (*Amblycorypha huasteca*) and a dead camel cricket (*Ceuthophilus* new
species), but they were ignored. The katydid accepted a live cricket. It was
caught up by the jaws alone and eaten after it had been in the cage for
ten minutes. One might suppose that the spiny, mantislike front legs
would be used to grab the prey, but they were not. Following capture by
the jaws the heels of the two front tibiae were brought together beneath
the prey to brace it and hold it in place. It required another ten minutes
for the predator to consume its meal entirely, legs, wings, and all. The following day we offered a second cricket. This one was snatched up less

Fig. 6-6
Greater arid-land katy-
did; male showing
aposematic wings

than two seconds after it landed on the cage floor. The predator stalked a
third cricket around the container, with several unsuccessful leaps before
the prey was caught with the jaws, a hunt ending after ten minutes. Consumption of the entire cricket, including legs, wings, and even the ovipositor, required fifteen minutes, and, as usual, the prey was eaten alive. The
forelegs were again used to brace the food but not in the manner one
might expect from spines that suggest raptorial limbs, except possibly for
the spines near the heel or tip of the tibiae that came in contact with the
prey. The katydid dropped several pieces but picked these up with the
jaws and consumed them after ingesting the greater portion. Some fallen
morsels were grazed upon later when the katydid returned after wandering
about.

One of the males ate an offered specimen of the superb cicada (*Tibicen
superba*) that was about half the size of the katydid. Even with prey this
large only the heels of the tibiae were used to brace it, not the tibial
spines. During the lengthy meal the predator chirped from time to time.

Days later it immediately accepted a piece of fresh apple, casting doubt on a label of strict carnivory. After one month this male ate a three-eyed coneheaded katydid (*Neoconocephalus triops*), a large and powerful jumper itself. When the carnivore leaped on it, the commotion was enough to knock both animals over onto their sides, where the predator lay for a few minutes before regaining its feet and resuming its meal. One week later it consumed three May beetles in the space of one hour. Several days later it ate a female cricket and another a day later.

Before eating its meal, the greater arid-land katydid often carries its prey about in its jaws for some time, much as a leopard does. It might be expected that the attack and the feeding process itself would be noisy considering the size of the animals involved and the hardness of the prey's exoskeleton. Nevertheless, both the capture and the consumption of the cricket were silent.

Females are rarely encountered. We dissected one specimen and found thirty-five eggs in varying stages of development. Most of these were brown and advanced in development. They are cigar-shaped, 6 mm long, and a little over 1 mm wide. It is not known if the females lay their eggs in the soil or in vegetation but Park Manager David Allen observed a female with her ovipositor inserted in the soil during the first week of August, and she appeared to be laying eggs.

Distribution: In the United States the greater arid-land katydid is restricted to central and western Texas and a small portion of eastern New Mexico and southern Arizona.

Remarks: Length = 75 mm. We began an e-mail dialogue with Theodore Cohn, who published the definitive work on this remarkable animal and its relatives thirty-six years before our own studies began. Cohn refers to the aggressive predator as "the red-eyed devil" because of the striking orange-red eyes with their false pupils and because of its tendency to bite hard when handled.

Similar species: The closely related species *Neobarrettia victoriae* might occur in the uplands bordering the swamps. Its eyes are white in life rather than red.

Blacklined shieldback katydid
Pediodectes nigromarginatus
(FIG. 6-7)

Biology: This is a predatory grasshopper that kills and eats juvenile stages of other grasshoppers as well as stink bugs and the pillbugs that are so abundant in the Ottine swamp. Like the greater arid-land katydid it is actually

Fig. 6-7
Blacklined shieldback katydid (*Pediodectes nigromarginatus*)

an animal of the adjacent upland. In captivity it eats flowers and flesh so that its diet, as well as the structure of its jaws, is similar to that of its more famous and destructive relative the Mormon cricket (actually a grasshopper too; Isely 1941, 1944). In northern Texas blacklined shield-backs aggregate beneath cactus and high aboveground in broad-leaved plants.

The wings are so small that neither sex can fly. Males use their forewings to stridulate when they court females, and when mating is completed, the females lay their eggs in soil. Their own wings, if present at all, might have no function. It has been said, and disputed, that the young nymphs do not escape by leaping, but by running.

Distribution: The unusual distribution of this animal illustrates a grassland association, for it occurs in a narrow band from extreme southern Texas to a northern limit in South Dakota. Within Texas this area can be described as a strip running north-south through the central part of the state.

Remarks: Length including ovipositor = 50 mm. Our single encounter with the blacklined shieldback occurred when we found a female "waiting" for us at a pasture gate only inches from the latch that opens into Soefje Swamp.

Similar species: None.

Fig. 6-8
Southern mole cricket (*Scapteriscus borellii*)

Southern mole cricket
Scapteriscus borellii
(FIG. 6-8)

Biology: The burrowing habit of the aptly named mole cricket is indicated by its strongly developed front legs, which suggest the front legs of a mammalian mole. Southern mole crickets are more carnivorous than related species and thus are more likely to become pests by tunneling through root systems than by eating them. The red imported fire ant, a fellow South American exotic, is thought to be among its prey. Both sexes fly, males sing to prospective mates from their burrows, and occasionally they show up at lights after dark. The song has been described as a "rich guttural 'grrr'" (Helfer 1953).

Distribution: This is an introduced animal that now ranges from the Atlantic Ocean to central Texas and by separate introduction, farther west into Arizona and California.

Remarks: Length = 35 mm. Southern mole crickets often feign death when captured. The common name is more fitting than one might imagine, for this exotic of the southern United States is now known to be native to southern South America instead.

Similar species: Other mole crickets occur in the Ottine swamps, but Texas populations of the southern species are recognizable by four pale spots on the otherwise dark region behind the head (Internet: University of Florida, Institute of Food and Agricultural Sciences, Mole Cricket Knowledgebase).

Fig. 6-9
Palmetto camel cricket (*Ceuthophilus* new species) from colony in stone tower

Palmetto camel cricket
Ceuthophilus new species
(FIG. 6-9)

Biology: We found the brown, humpbacked camel cricket (not yet formally known by a scientific name) only in, on, and near the sandstone water tower of Palmetto State Park. These unusual animals are nocturnal, wingless grasshoppers that cannot sing and presumably communicate by sense of touch via their long, filamentous antennae. They leave their hiding places in the tower just after dark, at approximately 9:15 P.M. in late May. Within minutes a dozen or so will be seen resting and crawling across the sandstone blocks. When touched, they escape with a powerful leap. We saw juveniles six feet aboveground in the foliage of an ash tree next to the building. Camel crickets are omnivorous, and peanut butter is a fine choice for those who wish to catch and raise them (Hubbell 1936). They form groups or aggregations by the attraction to a chemical pheromone perceived by the antennae (Nagel and Cade 1983).

Distribution: This new species of camel cricket is endemic to Texas and occurs nowhere in the world but in the central part of the state.

Remarks: Length = 25 mm. With the camera's flash the markings on this animal's body appear to have been painted on with gold leaf. A design on each side has been compared to the shape of a lyre, and the designs on its back are similar to those of a relative that were compared to a fleur-de-lis (Rehn 1907).

Similar species: Several camel cricket species might occur in the swamp. Their identification is difficult, and males are more easily identified than females. As of this writing the standard reference remains a huge tome from the early twentieth century (Hubbell 1936). We thank Ted Cohn for telling us that our species is new to science.

Short-tailed cricket
Anurogryllus arboreus
(FIG. 6-10)

Biology: The short-tailed cricket is unusual for living in family groups in burrows. These consist of a female and her offspring but no adult males. In fact, a male that attempts entry will be repelled. If he persists, the female bites off his leg and eats it (West and Alexander 1963). There are some remarkable similarities to ant nests here. For example, burrows consist of tunnels and chambers. Some are used for retreat from danger, others for

Fig. 6-10
Short-tailed cricket
(*Anurogryllus arboreus*)

the storage of food and waste. A mound of soil marks the entrance to the nest. The entrance is plugged from within against disturbance and during the time that the female has a brood. Eggs are laid inside the nest, carried to an egg pile, and groomed and guarded there. Some of the eggs, again like those of many ants, are "trophic" in function. They do not hatch and are eaten by the young crickets instead. Females do not allow their young to eat the viable eggs. Hatchlings, as many as fifty-six, remain in the nest for some time under the mother's protection, and she will advance against a nest disturbance rather than flee from it. The presence of nest symbionts, or "houseguests," is a further similarity to the ant colony. These small arthropods live beside, and even on, their hosts. We found the crickets only when they were attracted to lights after dark, and all individuals were males. Others have also seen this cricket abroad only after dark (Weaver and Sommers 1969). According to an authority on American crickets, the male's song is the loudest of all cricket songs in the United States (Hebard 1934).

A curious behavior is the habit of limited self-cannibalism or mutilation. They remove and eat their own hind wings within one day of becoming adults. Hence, these wings are not used for flight but only for food. The same fate befalls the rest of the female's body when she dies. Her young consume her remains before leaving the nest. Not believing the report of wing mutilation, we examined several specimens in the field and found that they did indeed lack the hind wings.

Distribution: From the Atlantic Ocean to a western limit somewhere in Texas, perhaps near the Ottine wetlands.

Remarks: Length = 17 mm. Short-tailed crickets are uncommon in the central Texas wetlands but reach pest status in eastern states when they consume cotton, potatoes, peas, strawberries, tobacco, and pine (Blatchley 1920). Vegetation that is not eaten is used to line the burrow.

Similar species: None.

Leather-colored bird grasshopper
Schistocerca alutacea
(FIG. 6-11)

Biology: This very large, beautiful yellow and black grasshopper does not favor grasses as much as the leaves of oak trees, and, perhaps not noted until now, the hoptree (*Ptelea trifoliata*), which it nearly defoliates when the nymphs are feeding in spring. Leather-colored grasshoppers prefer moist habitats such as marshes, bogs, seeps, and shrubby swamps (Helfer 1953; Hubbell 1960). The bright colors do not suggest leather in the least

Fig. 6-11
Leather-colored bird grasshopper (*Schistocerca alutacea*) laying eggs

and suggest instead some warning of a defensive capability perhaps acquired while feeding on toxic plants, but there is no reason to believe this upon consideration of their known foods.

Distribution: From the Atlantic Ocean to Arizona.

Remarks: Length = 50 mm. It is rare to identify populations with such high resolution, but the bird grasshoppers of the swamp appear to be intermediate forms resulting from hybridizations between the subspecies *lineata* and *albolineata*. The range of that form is limited to a small part of Texas that happens to have the Ottine wetlands at its center (Dirsh 1974).

Similar species: American bird grasshoppers (*S. americana*) do not occur in the Ottine swamps, though they are abundant in the Lost Pines not far northeast of these lowlands because of a preference for drier upland habitats.

Cattail toothpick grasshopper
Leptysma marginicollis
(FIG. 6-12)

Biology: This toothpick grasshopper, as the name suggests, is a slender species that clings to stems of cattails and giant cutgrass growing in marshes and other wet places. If approached, it sidles from side to side to remain hidden from view, much as a squirrel does on the trunk of a tree. We saw them on other plants, but they occur in greatest numbers on cattails in marshes far from trails. They begin mating on their host plant as early as April, and, remarkably for a grasshopper, eggs are laid inside the

Fig. 6-12
Cattail toothpick grasshopper (*Leptysma marginicollis*); mating pair

stems of aquatic plants (Preston-Mafham 1990). In Florida and Indiana one naturalist reported that they had never been seen on the ground, nor did they hop when disturbed. They always flew away instead (Blatchley 1920).

Blatchley found that only one population was known in the entire state of Indiana, though the species is common in southern states. This population lived at a pond that he visited again and again over the course of many years. "In 1893 and 1894 the insect was still present, though in rapidly decreasing numbers, as the pond was partially drained." Twenty-three years later Blatchley discovered that they were nearly gone: "In October, 1917, I again visited the former site of the pond, but found only a vast cornfield, with no signs of this or the other rare Orthoptera which formerly dwelt in numbers, in that locality" (Blatchley 1920, 308).

Distribution: From coast to coast within the United States where marshy conditions allow.

Remarks: Length = 38 mm. According to one source the cattail toothpick grasshopper is difficult to catch (Helfer 1953), but according to our own experience and that of Blatchley (1920) they can be taken with relative ease, even by hand.

Similar species: None.

Differential grasshopper
Melanoplus differentialis
(FIG. 6-13)

Biology: This is a large, yellow, very common, and very destructive crop pest that feeds on grasses and other herbaceous plants. It also scavenges the remains of fellow grasshoppers that have died. We saw them in summer in the cattail marsh of Palmetto State Park where the huge "green-eyed monster" robber fly (*Microstylum morosum*) preys upon the hoppers. The fly watches like a hawk from its perch on a twig. When a hopper takes to the air in front of an advancing boot, the predator pursues; but success might be rare, for we never saw them capture their prey. Mated differential females insert their abdomen into sandy soil and lay nearly two hundred eggs per clutch.

Distribution: From coast to coast within the United States.

Remarks: Length = 44 mm. When resting at night in the tops of herbaceous vegetation, the differential grasshopper remains motionless and may be plucked off the plant by hand. Fishermen take advantage of this by collecting buckets of the bait before a rising sun restores vitality to the creature's limbs. The adult differential grasshopper is particularly common in July.

Similar species: None.

Fig. 6-13
Differential grasshopper (*Melanoplus differentialis*); mating pair in North Soefje Marsh

Fig. 6-14
Two-striped
grasshopper (*Mer-
miria bivittata*)

Two-striped grasshopper
Mermiria bivittata
(FIG. 6-14)

Biology: When this big hopper is approached in a marshland, it soars a great
distance before landing once more on vegetation. In summer it often
perches at the tip of a tall stem during the hottest hours of the day.

Distribution: From the Atlantic Ocean to Nevada.

Remarks: Length = 40 mm. In other parts of the United States the two-
striped grasshopper occasionally becomes a pest (Helfer 1953).

Similar species: The large size in combination with the slanted face and
oddly thickened antennae help to distinguish this species from all others.
Confident identification requires the proper keys, because at least one
close relative might occur in the same area.

7

Cockroaches, Walkingsticks, Mantids, and Earwigs

Cockroaches are mostly omnivorous insects with a bad reputation that has been earned by some disease-spreading species that infest kitchens and bathrooms in urban settings. Three of these pests became established in the Ottine wetlands following accidental introduction from beyond the borders of the United States. From the technical point of view provided by the insect evolutionary tree, mantids are roaches too. They are roaches that abandoned omnivory for a predatory life. Earwigs, like mantids, are also closely related to roaches. The position in the family tree of the exclusively herbivorous walkingsticks, or "stick insects," has not been so clearly determined, but they are treated here because they bear an outward resemblance to the exclusively predatory mantids.

Smokybrown cockroach
Periplaneta fuliginosa
(FIG. 7-1)

Biology: In the swamp we saw the introduced smokybrown cockroaches well after dark and usually in or near buildings. Sometimes they were mating on the trunks of large trees near ground level. One of these sites is the large refectory built in the 1930s and now rented out for picnics. Another is an uninhabited water tower, showing that garbage is not a requirement for the roach's association with structures built by humans. It appears that this omnivorous roach has been naturalized in the wild as well, and we saw it preying upon pillbugs and helpless pupae of other insects. Dozens of individuals, including juveniles and females with egg cases, scrambled

Fig. 7-1
Smokybrown cockroach (*Periplaneta fuliginosa*), or palmetto bug; mating pair

from beneath the bark of a wind-thrown oak tree in Rutledge Swamp when their protection was pried away.

Distribution: Smokybrown cockroaches are probably natives of Asia, but commerce has introduced them into the United States, where they now occur from the Atlantic Ocean to a western limit in central Texas (Atkinson, Koehler, and Patterson 1991).

Remarks: Length = 33 mm. This species is appropriately known as the palmetto bug (Atkinson, Koehler, and Patterson 1991), and it does inhabit palms though we did not notice these roaches in the abundant dwarf palmettos of the Ottine wetlands. Palmetto bugs have the appearance of a household pest but more typically live out of doors near human habitations.

Similar species: Three large, closely related, and similarly introduced roaches of south-central Texas resemble the smokybrown cockroach. The most familiar of these is the American cockroach (*P. americana*; length = 34 mm), which is present in much smaller numbers than the palmetto bug and which differs by the very slender cerci (appendages) at the tip of the abdomen and by the much lighter coloration. The brown cockroach (*P. brunnea*) is more like the American species in its lighter color but like the smokybrown in its robust cerci. It is often misidentified as the palmetto bug.

Pennsylvania wood roach
Parcoblatta pensylvanica
(FIG. 7-2)

Biology: This species is a denizen of grasslands, forest, and wetlands. It is also one of the few native roaches that becomes a pest in homes. We usually encountered flying males at black lights after dark.

Distribution: From the Atlantic Ocean to San Antonio, Texas, not far west of the Ottine wetlands.

Fig. 7-2
Pennsylvania wood roach (*Parcoblatta pensylvanica*); male

Fig. 7-3
Female wood roach with egg case, possibly the banded wood roach (*Parcoblatta zebra*)

Remarks: Length = 25 mm. Pennsylvania wood roaches were considered rare in central Texas by one authority, but they are probably abundant.

Similar species: Several close relatives are likely to occur in these woods, such as the female shown in Fig. 7-3, which might be the banded wood roach (*P. zebra;* length = 16 mm). Confident identification requires microscopic examination and the proper keys.

Fig. 7-4
Cuban green cockroach (*Panchlora nivea*)

Cuban green cockroach
Panchlora nivea
(FIG. 7-4)

Biology: The beautiful green color of this introduced cockroach seems to contradict the image of its kind as dirty, dark, noxious animals. It no doubt provides effective camouflage for a life spent in lush green vegetation, including the leaves of bananas in tropical climes (Hebard 1943). We never saw the species on living plants. Instead, we found them in rotting pecan logs and, beginning in April, at black lights after dark. Their presence in pecan logs is a recent invasion of an ecological niche. They are very fast on their feet, and escape is a combination of racing, fluttering, and flying once the protection of cover is pried away. We found both sexes in the swamp, further supporting the hypothesis that they have become established here. The diet is unknown though presumably omnivorous like that of other cockroaches.

Distribution: The Cuban green cockroach is a tropical species that arrives periodically in ports around the world in banana shipments from the

Fig. 7-5
Pale-bordered cock-
roach (*Pseudomops
septentrionalis*);
mating pair

Caribbean. It has now established domicile in the United States along the
Gulf Coast from Florida to Texas, but we can find no previous reports of
wild U.S. populations with the possible exception of Brownsville, Texas
(Hebard 1943). According to the most recent assessment, "its habits suggest that it may occur in some natural communities, such as mesic ham-
mocks" (Atkinson, Koehler, and Patterson 1991, 72). We discovered the
Ottine swamp population in just such a habitat, with an even more precise preference for decomposing wood beneath the bark of pecan logs.
Remarks: Length = 18 mm. Strangely, we did not see the Cuban green cockroach around the heavily trafficked sandstone refectory building, where
people often have large picnics and outdoor banquets, unless we were running black lights after dark among the surrounding trees. The Cuban
green cockroach appears from time to time at porch lights in the nearby
city of Austin.
Similar species: None.

Pale-bordered cockroach
Pseudomops septentrionalis
(FIG. 7-5)

Biology: This cockroach is perhaps even more beautiful than the Cuban
green cockroach. Its coloration is an attractive mixture of dark, light, and
red. Our experience confirms the generalization reported by others that it
tends, almost without exception, to dwell in low-growing herbaceous
plants. Sometimes we saw pale-bordered roaches beneath boards or on

tree trunks, and most of our sightings occurred during daylight hours. They were not attracted to lights after dark.

Pale-bordered roaches seem to prefer low areas near bodies of water, because the few specimens we saw were in a dry creek bed in Austin, along pond margins in the Lost Pines forest of central Texas, only a few inches from a creek in the outback of the Ottine swamp, and in frostweed growing near the lagoons of Palmetto State Park. Almost nothing else is known about this creature's biology.

Distribution: In the United States the pale-bordered cockroach occurs only in Texas, from near the Louisiana border in the east to the Devils River area in the west (Atkinson, Koehler, and Patterson 1991).

Remarks: Length = 15 mm. Bright colors suggest a tropical exotic like the green Cuban species, but this elegant animal is a Texas native first made known to science from specimens collected in Brownsville.

Similar species: None.

Fig. 7-6
Dark wood roach
(*Ischnoptera deropeltiformis*)

Dark wood roach
Ischnoptera deropeltiformis
(FIG. 7-6)

Biology: Dark wood roaches are primarily ground dwellers that frequent leaf litter and logs. Males are unusual among cockroaches because they fly rather than run when disturbed (Blatchley 1920). Females have reduced wings and cannot fly under any circumstances.

Distribution: From the Atlantic Ocean to a western limit in central Texas.

Remarks: Length = 18 mm. Dark wood roaches are said to prefer moist forests, and if so, they have a fine home in the Ottine swamps.

Similar species: Dark color and slender form combined with orange legs will distinguish this roach from all others.

Giant walkingstick
Megaphasma dentricus
(FIGS. 7-7 – 7-8)

Biology: We first encountered this huge insect at the end of May when a female fell from the heights of a towering ash tree in the Ottine swamp. It was large enough to strike the understory vegetation with the sound of a broken twig. One week later a second female was found on the foliage of a boxelder tree at a height of six feet. In confinement it released several small brown eggs that resembled seeds. After dark we found two males, one on a palmetto frond and another on the trunk of an ash tree.

Fig. 7-7 Giant walkingstick (*Megaphasma dentricus*); male on palmetto frond

Fig. 7-8 Giant walkingstick; female on stone tower

This is a leaf-eating vegetarian, and previous records of its host plants include grapevines in the river bottoms of nearby Victoria, Texas (Caudell 1903), post oaks in the same area (Hebard 1943), and "shrubs and trees" (Somes 1916). Thus, the literature does not seem to support the idea that the giant walkingstick is entirely limited to the heights of trees where it will seldom be seen. Nevertheless, in Texas the species is "extremely local" in its appearance (Hebard 1943). Hebard and a colleague visited the site in Victoria where the walkingstick had been reported from river bottom grapevines, but a search of many hours ended in frustration without a single specimen. Hebard suggested that nocturnal collecting with a flashlight or headlamp might meet with greater success.

Distribution: The giant walkingstick has an unusual inland range from New Mexico in the west to Kentucky in the east.

Remarks: Reaching lengths of nearly 180 mm, this animal is the longest insect in the United States. An early authority described it as more tropical in appearance than any of the other stick insects of this country and noted that "the large size commands attention wherever seen" (Caudell 1903).

Similar species: None.

Carolina mantis
Stagmomantis carolina
(FIG. 7-9)

Biology: Though we never saw a Carolina mantis under natural conditions, it is probably abundant in the strict sense of occurring in large numbers. The species is nearly invisible due to an apparent preference for the heights of trees rather than the open lowlands of meadows. Occasionally a male Carolina mantis showed up at our black lights, but we never saw the flightless female.

Carolina mantids are predators with a broad diet including flies, roaches, and other insects. In nearby Austin a male flew to a hanging hummingbird feeder and awaited the arrival of bees, but the females must walk or leap wherever they go because their wings are too small for flight. The egg cases that they leave on bark, wooden fence posts, and even on metal poles are instantly recognizable as tough brown pods with a braided appearance. In fact, these are more likely to be seen than the insects themselves.

Distribution: From the Atlantic Ocean to the Rocky Mountains and Arizona.

Remarks: Length = 57 mm. Common names for mantids include soothsay-

Fig. 7-9
Carolina mantis
(*Stagmomantis carolina*)

ers, mule killers, devil's riding horses, and in the southwestern United States and Mexico, campomoche.

Similar species: Carolina mantids resemble the more famous praying mantis, but the latter does not occur in the wetlands. This is the only mantid that we saw.

Yellow-winged earwig
Vostox brunneipennis
(FIG. 7-10)

Biology: This is a wild, rarely encountered earwig with two yellow spots on its back and a biology that remains unknown except for its habitat, which seems invariably to be underneath the bark of decomposing logs. We found them in a large, decaying pecan log far from the trails and in groups beneath the bark of a wind-thrown oak in Rutledge Swamp. They run fast and are difficult to capture because they are small and flat. This leads to frustration when one attempts to extract them on or beneath bark. In the early part of the twentieth century one of North America's experts on earwigs reported that the yellow-winged species had been found only on magnolia and birch and under the bark of willow trees (Hebard 1917, 1943).

Distribution: Apparently occurring from the Atlantic Ocean to a western limit in central Texas not far from the Ottine swamp, this species "is so rarely encountered that the records give little definite information as to the limits of its distribution" (Hebard 1917, 315). According to this same

Fig. 7-10
Yellow-winged earwig (*Vostox brunneipennis*) from pecan log

article, the activities of a nineteenth-century collector suggest that it might be more common in central Texas than elsewhere within its range. A second source reports the species from Arizona but agrees with Hebard that it is uncommon (Helfer 1953).

Remarks: Length = 12 mm. During the courtship ritual of a closely related species, the male grips the female with the forceps at the tip of the abdomen and hurls her through the air for a distance of an entire body length (Briceño and Eberhard 1995).

Similar species: None.

8
Wasps and Ants

Wasps and ants are among the most conspicuous invertebrates in the Ottine wetlands. Airborne red and black paper-making wasps are reliable sights in spring, summer, and fall, and on the ground red imported fire ant mounds grow to enormous size in wooded swamplands that resemble the exotic pest's South American floodplain home more than any other ecosystem in Texas. From both species as well as from many others, this red color is a warning, a reminder to us all that wasps, ants, and bees are the only insects that sting.

Long-tailed boxelder wasp
Megarhyssa macrurus
(FIGS. 8-1–8-2)

Biology: Females insert 75 mm long ovipositors into tree trunks and lay their eggs inside. A parasitic grub hatches from the egg and consumes the larva of the pigeon tremex (*Tremex columba;* see Fig. 8-3). Adult males emerge after completing their development and compete to mate with emerging females. When the latter chew their way to the surface, the resulting noise sounds like "a person eating a raw carrot" (Gibbons 1979, 722).

We watched these wasps in mid-April as females laid eggs in dead portions of boxelder trunks. Males flew from tree to tree and landed near the females, but they never attempted to mate. Birds prey upon "long-stings," and egg-laying females are particularly vulnerable because their ovipositors cannot be removed from the wood without expending much time and effort. In 1880, one naturalist came upon a scene of attrition in which the helpless animals had been torn away from their anchors, perhaps by birds.

Fig. 8-1
Long-tailed boxelder
wasp (*Megarhyssa
macrurus*); female
laying eggs in mori-
bund boxelder

Fig. 8-2
Long-tailed boxelder
wasp; male

He saw "sticking out of the bark many ovipositors which had belonged to unfortunate visitors of the previous summer" (Harrington 1882, 84).

Distribution: From the Atlantic Ocean to Arizona.

Remarks: Length of females = 120 mm, including the long egg-laying tube. A long history of confusion and disagreement surrounds this wasp. Some early American entomologists believed that the grub stage feeds upon wood and is not parasitic on another animal. The egg-laying process was also described as drilling or sawing rather than inserting. These wasps are ideal subjects for observation because they seem to ignore their surroundings and are unlikely to fly off unless they are touched. We saw them only during a brief period in spring.

Similar species: None.

Fig. 8-3
Pigeon tremex
(*Tremex columba*)
laying eggs on hackberry tree

Pigeon tremex
Tremex columba
(FIG. 8-3)

Biology: Red and black adult females lay their eggs in the wood of boxelder, elm, oak, sycamore and, as we discovered, in smooth hackberry (*Celtis laevigata*). This particular individual chose the trunk of a large living tree at a location five feet aboveground. The egg-laying structure (ovipositor) pierces the wood to depths of 15 mm or more, but the holes are not evident because they are a mere 0.2 mm wide (Stillwell 1967). White larvae hatch from the eggs and grow to lengths approaching 50 mm. Presumably this grub stage subsists on a diet of fungus that is consumed along with the wood it infects. The adult stage is reached in the second year of development, and females mate quickly upon emergence from the tree, laying their own eggs soon after.

The most remarkable feature of the pigeon tremex life cycle is its symbiotic relationship with the fungus *Daedalea unicolor*. Adult females carry this fungus inside compartments located between segments of the abdomen. When eggs are laid, the animal injects the fungus via the long, piercing ovipositor and via the eggs themselves, which carry a cap of fungus into the wood. Thus, they bring the seed of their own garden with them if it is true that the grubs graze on the symbiont. Soon the larva will harbor the fungus within its own set of specialized organs. When the adult female emerges from its last juvenile stage, it uses its ovipositor like a straw to draw the fungus into its body from the shed skin, thus completing the cycle for both insect and fungus.

An enemy of the pigeon tremex is the long-tailed wasp (*Megarhyssa macrurus*) that we saw inserting its own ovipositor into boxelder trunks in search of the tremex grub (Fig. 8-1). In this case the egg is laid near the host, and upon hatching the parasitic grub climbs aboard and begins eating its victim.

Distribution: From coast to coast within the United States.

Remarks: Length = 47 mm. The red and black coloration suggests mimicry of red and black paper wasps that inflict a painful sting.

Similar species: Lack of a slender "wasp waist" and the cylindrical shape of the abdomen readily distinguish the harmless pigeon tremex, a species of sawfly, from the red wasps.

American elm sawfly
Cimbex americana
(FIG. 8-4)

Biology: We found a single black-striped caterpillar-like larva on an iris leaf where it presumably fell from an elm tree overhead. These caterpillars feed on the foliage of elms and apparently willows too (Wong 1954).

Distribution: Widespread in the United States and Canada.

Remarks: Length = 25 mm. We never saw the adult stage despite the abundance of elms and willows in these wetlands.

Similar species: Though the elm sawfly is a wasp rather than a fly, it does bear some resemblance to the mydas fly (*Mydas clavata;* Fig. 5-22).

Paper wasps
Polistes exclamans, P. metricus
(FIGS. 8-5–8-6)

Biology: Paper wasps build open-faced nest combs with a paper made from chewed wood and saliva. The larval stages living inside the cells are fed with caterpillar flesh that has been previously chewed by foraging adults

Fig. 8-4
American elm sawfly
(*Cimbex americana;*
right) and Mydas fly
(*Mydas clavata;* left)

Fig. 8-5
Guinea wasp
(*Polistes exclamans*)
on its nest at the
refectory building in
Palmetto State Park

Fig. 8-6
Red wasp (*Polistes
metricus*) with prey
in its jaws, apparently
the remains of a
caterpillar

that find their prey on foliage. Both species treated here will deliver a painful sting if the nest is approached, particularly the smaller and lighter-colored Guinea wasp (*P. exclamans*). The result is a sharp pain that weakens quickly and lingers throughout the day as a dull ache.

Distribution: Guinea wasps occur from coast to coast within the United States, whereas the red wasp (*P. metricus*) occurs from the Atlantic Ocean to Texas.

Remarks: Length of *P. exclamans* = 15 mm. Other species of this genus in the area, including the red wasp, tend to be larger, at around 23 mm or so. We noticed that paper wasps commonly attach their nests to tree trunks in wilder areas, particularly the boxelder, so that the comb is exposed to the elements. As a result one of us was stung several times without warning.

Similar species: The Carolina red wasp (*P. carolina*) is also common in central Texas. It resembles *P. metricus*, but its color is brighter red.

Klug's cowkiller
Dasymutilla klugii
(FIG. 8-7)

Biology: The hairy, wingless female races over sandy soil in search of an insect's nest in which to lay her eggs. It is believed that Klug's cowkiller invades the burrows of the huge cicada-killer wasp. An egg is laid near the offspring of the killed, ejected, or outmaneuvered host, and when the larva hatches, it consumes the resident grub. Though the usurper is much smaller than the cicada-killer, she has a panoply of weapons that make up for the difference in size. These include a huge and highly mobile sting, a defensive chemical secretion, great speed, and a tough, slippery exoskeleton (Schmidt and Blum 1977). Red and black warning colors combined with an ability to squeak by rubbing body parts together help ward off potential enemies that might interfere with her search aboveground. Male cowkillers have wings and spend much of their time flying low over the sand in search of mates. Sometimes they are seen at flowers. They often fly to black lights at night.

Distribution: This is one of the few western insects of the region, occurring from Arizona to an eastern limit near the Ottine swamps.

Remarks: Length = 20 mm. Klug's cowkillers are active, attractive animals with a fuzzy, endearing look, but females should never be handled because they received their common name in emphasis of a very painful sting. The pain of these wasps' stings among wasps of the United States is exceeded only by that of the stings of certain giant tarantula-hunting species. Nev-

Fig. 8-7
Klug's cowkiller
(*Dasymutilla klugii;*
left) and cicada-killer
wasp (*Sphecius spe-
ciosus;* right)

ertheless, the common name of cowkiller overstates the sting's potency.
When caught in net or hand, the male also makes stinging motions with
its abdomen, but his bluff is of course unsuccessful because no male insect
can sting.

Similar species: The large size, in combination with the particular arrange-
ment of red and black colors, distinguishes this common cowkiller from
other species.

Cicada-killer wasp
Sphecius speciosus
(FIG. 8-7)

Biology: The huge female hunts cicadas in the boughs of trees, paralyzes
them with a sting, and flies back to her burrow in the sand with the prey
slung upside down beneath her body, much like a papoose that humans
carry on the chest. Nests have one primary tunnel, several secondary
branches, and a cell at the end of each branch that is stocked with one or
more cicadas. The wasp lays an egg, and when the grub hatches, it eats
the living but helpless cicada. At the time of writing it was discovered that
the female excavates the burrow with the help of large spurs on its legs
(Coelho 2002).

Distribution: From the Atlantic Ocean to the Rocky Mountains.

Remarks: Length = 38 mm. This is the largest wasp in the wetlands as far
as we know. Reports differ as to the severity of its sting. Humans rarely
feel its effects in any event because solitary wasps are less likely to use
this defense than social wasps, such as the easily irritated yellowjackets
and red wasps.

Similar species: None.

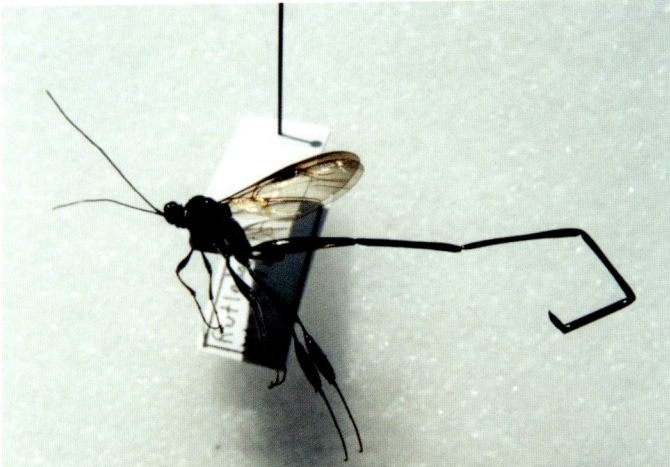

Fig. 8-8
Black June beetle wasp (*Pelecinus polyturator*)

Fig. 8-9
Striped June beetle wasp (*Myzinum quinquecinctum*) in sleeping aggregation on wax myrtle

Black June beetle wasp
Pelecinus polyturator
(FIG. 8-8)

Biology: The adult female is occasionally seen in wooded areas, such as that where we found ours, but males are rare outside the Tropics. In Texas reproduction probably occurs routinely without sex. Unmated females use their long and distinctively shaped abdomen to search in the soil for June beetle grubs, which are parasitized and eventually killed by larvae after these hatch from their eggs (Johnson and Musetti 1999).

Fig. 8-10
Ichneumon wasp
(*Therion circum-
flexum*) with remains
of its moth host's
body and cocoons

Distribution: From the Atlantic Ocean to New Mexico. In Texas the wasp is
known to occur only in the southeastern sector.

Remarks: Length of females = 70 mm, including the long egg-laying tube.
Because of the female's size, the abdomen, so suitable for finding grubs in
the ground, is a burden during the mating process and in the air, so her
flight is slow and awkward (Brues 1928). The wasps occasionally sting
when handled, but the result is not serious.

Similar species: None. This wasp is rarely collected (Aguiar 1997); there are
only two additional species in the entire family, and neither of those two
occurs in Texas. Black and yellow June beetle wasps of the genus *Myzinum*
parasitize the same hosts, and the adults nectar at a wide variety of flowers, including the deadly hemlock of the deep swamp itself (Krombein
1938). We encountered swarms of males in summer as they convened to
form sleeping aggregations on wax myrtle leaves near Rutledge Creek (Fig.
8-9) and on pecan leaves at the border of North Soefje Marsh. These are
M. quinquecinctum (length = 24 mm) and *M. berlyi* (length = 22 mm),
respectively.

Ichneumon wasp
Therion circumflexum
(FIG. 8-10)

Biology: In the grub stage this wasp lives parasitically inside the body of
moth caterpillars. Adult females require less than one second to inject an

egg into the host with their ovipositor or "sting." The stage hatching from the egg lives within a sac, and its protruding jaws nibble at the host's surrounding tissues (Slobodchikoff 1973, 1974). Several moth species serve as hosts, including the American dagger moth *Acronicta americana,* which we believe were the hosts of the two male wasps that we cultured to the adult stage. We collected the cocoons of these moths beneath the bark of a willow tree adjacent to a cattail marsh.

Distribution: The distribution map shows a coast-to-coast occurrence but with no previous record from Texas (Slobodchikoff 1977).

Remarks: Length = 23 mm. If geographic distribution is consulted when using the identification key, one might be swayed toward an identification as *T. fuscipennis* (Slobodchikoff 1977).

Similar species: This wasp is a member of the family Ichneumonidae. It is a large group, and it is notoriously difficult to identify the species.

Slender ponerine ant
Leptogenys elongata
(FIG. 8-11)

Biology: These sleek brown predators make their nests beneath logs and stones and within decomposing pecan logs. The highly specialized diet consists of the small and very abundant terrestrial crustaceans known as pillbugs. "The earth surrounding the entrances to the nests is invariably white with innumerable bleaching limbs and segments of the crustaceans, showing that great numbers of these animals must be habitually destroyed

Fig. 8-11
Slender ponerine ants (*Leptogenys elongata*) with cocoons

Fig. 8-12
Enormous mound of the red imported fire ant (*Solenopsis invicta*) in South Soefje Swamp

by the ants" (Wheeler 1904, 253). Slender ponerine ants are also notable for the absence of a winged queen. The egg layers look much like ordinary workers instead. Colonies are small and consist of fewer than one hundred individuals.

Distribution: In the United States this species might be restricted to Texas, though it likely also occurs in Louisiana.

Remarks: Length = 6 mm. Nests are encountered from time to time but are not as apparent as one might expect given the abundance of pillbugs in these lowlands.

Similar species: None.

Red imported fire ant
Solenopsis invicta
(FIG. 8-12)

Biology: The red imported fire ant is an exotic species from South America that builds large mounds on natural floodplains and in unnaturally disturbed areas. This is an aggressive animal that erupts from the nest by the hundreds or thousands to sting all intruders. Stings are painful and are usually followed by a white pustule within twenty-four hours.

The fire ant is omnivorous. It eats animals and plants whether living or dead, as well as secretions and excretions of the same. Winged males and winged queens fly from the nest after rains, mate in the air, and return to earth. The queens dig a new nest, and the males die.

Nests in the Ottine swamps grow to enormous sizes at the interface between pasture uplands and creek and river bottoms because these habitats are remarkably similar to the ant's primal home in the Paraguay River floodplain of South America. The two extremes of very dry and very wet are avoided, so swamp-dwelling fire ants are more common in logs than in the soil itself.

Distribution: Within the United States the red imported fire ant occurs from the Atlantic Ocean to central Texas, with spot infestations farther north and west. A major infestation was developing in southern California at the time of this writing, as well as an invasion of the Australian continent and, in late 2004, Taiwan.

Remarks: Length = 5 mm. The red imported fire ant invaded the Gulf Coast of the United States in the early years of the twentieth century (Taber 2000). This introduced species is now one of the most apparent animals in the wetlands.

Similar species: The tropical fire ant (*S. geminata*) might occur in the swamps, marshes, or uplands, but we never saw it. The largest workers have unusually big heads.

Pennsylvania carpenter ant
Camponotus pennsylvanicus
(FIG. 8-13)

Biology: The worker caste ranges from small individuals, or "minors," to "majors" that might be the largest worker ants in the wetlands. They live in trees, buildings, and other structures and are omnivorous feeders that kill and scavenge. After dark in Palmetto State Park dead, weak, and vulnerable insects are quickly consumed or overpowered by large numbers of Pennsylvania carpenter ants.

Distribution: From the Atlantic Ocean to a western limit in central Texas near San Antonio.

Remarks: Length = 12 mm. Luckily, these large and active ants are incapable of stinging, but they do bite. In some parts of the eastern United States carpenter ants earn their common name not by construction but by destruction, and thus they become pests of timber in homes and buildings.

Fig. 8-13
Pennsylvania carpenter ants (*Campono-tus pennsylvanicus*) devouring a true katydid

Fig. 8-14
Acrobatic ants (*Crematogaster lineolata*) on decaying log

Similar species: The large size and black color should allow for no confusion with any other species.

Acrobatic ants
Crematogaster lineolata
(FIG. 8-14)

Biology: These red and black omnivorous ants are members of the "stinging" group that includes fire ants and harvester ants. Yet their stinger is not so useful as a barb, being replaced as a defense by chemicals that are secreted through it. They live in and perhaps under decomposing snags and logs (such as willow and ash) rather than live exclusively in the soil and are especially conspicuous when foraging after dark for living or dead insects. This particular species was reported from dry habitats in the south but not from marshy habitats like those of wetlands (Johnson 1988).

Distribution: From the Atlantic Ocean to a western limit in the Rocky Mountain region.

Remarks: Length = 5 mm. The common name acrobatic ant is derived from a habit of raising and curling the back portion of the abdomen into the air when the animal is disturbed.

Similar species: Only a very few acrobatic ant species occur in central Texas, but they are surprisingly difficult to identify, partly because the group has not been adequately studied.

9

Nerve-winged Insects, Scorpionflies, and Hangingflies

The nerve-winged insects are among the most primitive of all insects that pass through complete metamorphosis. A complete metamorphosis, exemplified by the familiar example of the butterfly, is one in which the animal hatches as a larva from an egg and develops through a pupal stage before reaching maturity as an adult. Dobsonfly wings in particular retain the impress of an ancient condition that has gone through much modification in other groups, such as beetles, butterflies, and wasps. Scorpionflies and hangingflies are remarkable, enigmatic creatures that have been identified as close relatives of the true flies or even as the closest living relatives of the parasitic, wingless fleas.

Texas dobsonfly
Corydalus texanus
(FIG. 9-1)

Biology: The monstrous adults are seldom seen unless one happens to be near an outdoor light after dark, at the right time of year, and close to a stream or river. The male has a pair of long jaws that resemble ice tongs. Females have shorter jaws that more closely resemble wire cutters. The predatory larval stage, known as the hellgrammite to fishermen who prize it as bait, lives in flowing waters and resembles a wingless adult female. Adults do not live long and probably do not have to feed at all, though they accept sugar solutions under laboratory conditions.

Males use the big mandibles to court females and to fight competing males. These battles are fierce enough to break the jaws like jousting lances. Once mated, the female lays a mass of eggs on vegetation over-

Fig. 9-1
Texas dobsonflies
(*Corydalus texanus*)
courting at a black
light above the San
Marcos River

hanging water, and upon hatching the young hellgrammite falls into its aquatic home. When the time comes to transform to the adult stage, it crawls out of the water and pupates in a cell beneath a rock or log (Contreras-Ramos 1998).

Distribution: The Texas dobsonfly is one of the few western insects occurring in the swamps and was previously known from California to Laredo. Our record from Ottine thus marks a new eastern record for the species.

Remarks: Length of male = 70 mm. Looks can be deceiving, for the ice-tong jaws of the male are not as formidable as the short, powerful pincers of the female. When we allowed males to bite our fingers, the result was never more bothersome than a quick painful pinch that often did more damage to the insect than to us, for the tip of one jaw sometimes breaks off in the process. On the other hand, hellgrammites do bite effectively as is well known among fishermen, and we had no desire to confirm this by allowing the similarly equipped adult female to do the same.

This animal, especially the male, never fails to arouse interest in those who cross its path. While we were examining specimens in the swamp, a Texas Parks and Wildlife official approached us wanting to know the name of a strange bug he had just seen. It was the same dobsonfly species that we were ourselves examining.

Similar species: Two additional species might occur in the Ottine swamp. The yellow dobsonfly (*C. luteus*) is easily distinguished because it has no distinct color pattern on top of the head. The Texas dobsonfly (*C. texanus*) is distinguished from the eastern species (*C. cornutus*) by the dark ring surrounding each, or at least most, of the pale dots that are scattered across

Fig. 9-2
Fishfly (*Chauliodes rastricornis*)

the wings. Eastern dobsonflies have the pale dot but not the dark ring surrounding it.

Fishfly
Chauliodes rastricornis
(FIG. 9-2)

Biology: Fishfly biology is much like that of the dobsonfly (Dolin and Tarter 1981). Fishflies differ in their development because the juvenile stage lives in the quiet waters of the Ottine swamps rather than in the swift waters of the San Marcos River. They also differ in their smaller size and in the lack of outsized mandibles in the male sex. Adults die not long after transforming from the pupal stage, and they probably do not have to eat. Nevertheless, one female was found at our black light with her mandibles buried in banana bait. This might be the first record of such an observation for this species.

Distribution: From the Atlantic Ocean to a western limit, in the southern United States at least, in the region of the Ottine swamps.

Remarks: Wingspan = 80 mm. The fishfly's jaws are not as strong as those of the larger dobsonfly, and fishflies cannot bite effectively if handled. At a regular blacklighting spot on a bank of the San Marcos River, we found fishflies arriving earlier in spring than dobsonflies (at the beginning of April), and they stopped coming by the end of the month, when dobsonflies still crawled across the illuminated white bedsheet. Even alongside the still waters within their own swamp fishflies stopped coming to lights before the end of May.

Similar species: A closely related fishfly might occur in the swamp. That species is *C. pectinicornis,* and it can be distinguished by a yellow color pattern at the back of its head in place of the brown pattern of *C. rastricornis.*

Fig. 9-3
Alderfly (*Sialis velata*)
Photo courtesy of Jason Locklin.

Alderflies
Sialis itasca and relatives
(FIG. 9-3)

Biology: Alderflies are small, black, and rather uncommon relatives of the huge dobsonflies. They are said to be slow and awkward diurnal fliers (Chandler 1956), though we found it challenging to net them from the air as they flew by. This species is known to occur in both rivers and lakes (Ross 1937). Females lay their eggs on vegetation overhanging the water, and when larvae hatch, they drop down to make their home on the bottom as gilled predators of other insects. When the time arrives to transform to the adult stage, they crawl ashore and pupate inside earthen cells underground.

We captured our specimens of *S. itasca* in flight one by one on the afternoon of March 31 along the shore of the oxbow lake of Palmetto State Park, at a time of day when alderflies are known to be most active. We found them flying ashore at one spot only, and this brought to mind the flyways used by migrating birds. The flight was sampled between 1 P.M. and 2:30 P.M. in bright light and warm temperature. This is one day earlier than the April 1 date given as the beginning of the adult's emergence period in nearby Brazos County (Whiting 1991). Adults have been seen as late as September 30 in Indiana. Some of the alderflies we watched descended to shoreline vegetation and rested on branches overhanging the shore.

Distribution: From the Atlantic Ocean to Texas, though this appears to be the first record from Gonzales County, which might represent the westernmost record for the species.

Remarks: Length = 10 mm. Relatively little information is available for this alderfly, even when compared to the paucity of knowledge concerning its better-studied relatives (Canterbury 1979). The common name refers to a habit of resting on leaves of alder trees.

Similar species: *Sialis americana* and *S. velata* are the only other alderflies known to occur in central Texas. These insects are exceedingly difficult to identify. We thank John Oswald for determining the species, and we thank Jason Locklin for providing us with the results of his research on the alderflies of central Texas.

Macleay's owlfly
Ululodes macleayana
(FIG. 9-4)

Biology: We found no information on the biology of this odd-looking predatory creature that arrived infrequently at our black lights after dark. Based upon the few observations that have been made upon related species in the United States, adults probably fly close to the ground near dark in search of caddisflies, mayflies, beetles, moths, and true flies (Halstead 1989). When resting on twigs, they hold the body away from the wood surface at nearly a right angle and thus appear to be twigs themselves. The larva is a monstrous-looking predator that covers itself with debris and kills its prey with large, venomous jaws.

Distribution: From the Atlantic Ocean to a western limit in Texas.

Remarks: Length = 25 mm. Surprisingly little is known about owlflies. We encountered them only at our black lights after dark and never under conditions that might be described as natural. Even more remarkable is the experience of another investigator who saw them under natural conditions but was never able to attract them to artificial lights (Halstead 1989).

Similar species: Six species are recognized in the United States according to

Fig. 9-4
Macleay's owlfly
(*Ululodes
macleayana*)

a key in manuscript form made available to us at the time of writing, and Texas is the only state in which all six occur. Unfortunately, one of these could not yet be reliably distinguished from the rest, and this happens to be the only owlfly that might be endemic to south-central Texas wherein lies the Ottine swamp. Our tentative identification is based upon examination of the specimens by John Oswald of Texas A&M University.

Mantisflies
Mantispa viridis, M. pulchella
(FIGS. 9-5–9-6)

Biology: Adult mantisflies earn their common name by a remarkable similarity to praying mantids. The latter are much larger, and the two insect groups are not close relatives at all, despite front legs specialized in like manner for predation. Adults of both sexes are winged and are good fliers. They feed upon house flies and other insects, whereas the life of the larva seems nearly incredible.

After hatching from an egg at the tip of a long stalk, the green mantisfly grubs (*M. viridis*) scramble about in search of a spider's egg sac. Cutting their way inside, they feed upon the eggs until they are nearly adults. Then they chew their way back out of the sac, make a final molt to the winged condition, and seek out mates to begin the next generation (Brushwein, Hoffman, and Culin 1992; Brushwein, Culin, and Hoffman 1995a, 1995b; Rice and Peck 1991). Spider hosts occurring in the Ottine wetlands include black widows, wolf spiders, and crab spiders.

Fig. 9-5
Green mantisfly
(*Mantispa viridis*)

Fig. 9-6
Beautiful mantisfly
(*Mantispa pulchella*)

Courtship is rarely observed except under laboratory conditions. It begins when a male and female face one another while flexing the predatory front legs in a behavior known as sparring. The male has a series of glands between plates on its back, and it protrudes these, fanning its wings as if to send a chemical message toward the potential mate. The message might make the female more receptive, or it might simply reduce the probability that she will eat the male rather than mate with him. More sparring follows copulation, and soon the female will lay her eggs, perhaps on trees or under eaves where spiders are abundant.

The beautiful mantisfly (*M. pulchella*) differs in its development by its ability to board spiders as a parasite in the larval stage so that it is not entirely restricted to feeding on eggs within a sac (Brushwein, Culin, and Hoffman 1995a).

Distribution: Green mantisflies occur from the Atlantic Ocean to at least as far west as central Texas and south into Central America. The beautiful mantisfly occurs from the Atlantic Ocean to at least as far west as the Rocky Mountains.

Remarks: Length of *M. viridis* = 13 mm; length of *M. pulchella* = 8 mm. As interesting as these animals are, they are unlikely to be seen because in our experience the adults appear only at lights after dark, whereas the grub or larval stage lives within the egg sacs of spiders.

Similar species: Close relatives probably occur in the area. The green mantisfly is readily identified by its color, but identification of other mantisflies should be made by a specialist, for which we thank John Oswald.

Fig. 9-7
Scorpionfly (*Panorpa nuptialis*) scavenging the remains of a grasshopper

Scorpionfly
Panorpa nuptialis
(FIG. 9-7)

Biology: Adults are odd-looking red insects with long beaks and black and orange wings. They appear in fall and winter in small numbers to scavenge dead animals and sip nectar from flowers. We saw them eating dead grasshoppers and smashed snails. Favored habitats are said to be open areas along tree lines and power line cuts, but we found many individuals and mating pairs in the understory near the oxbow lake in Palmetto State Park.

Male scorpionflies bear a harmless structure at the tip of the abdomen that resembles a scorpion's stinger. This is used to grasp the female as the two prepare to mate. He also vibrates his wings and presents her with a gift in the form of a salivary secretion. Both sexes regurgitate a brownish substance when handled. They are clumsy fliers, and when disturbed, they set down again after only a few seconds in the air. The bright colors of the body and the wings warn potential enemies of distastefulness, or perhaps they only mimic animals that truly do taste bad or deliver a painful sting. Scorpionfly larvae resemble caterpillars. However, they do not live in vegetation or feed upon green leaves. They crawl in the soil and scavenge for food.

Distribution: From the Atlantic Ocean to the Austin, Texas, area near the Ottine wetlands.

Remarks: Wingspan = 36 mm. According to one report scorpionflies are becoming scarcer because red imported fire ants prey on the vulnerable soil-dwelling larval stage.

Similar species: None.

Hangingfly
Bittacus punctiger
(FIG. 9-8)

Biology: Hangingflies are some of the most interesting insects to be found anywhere. Their scarcity only adds to the wonder provoked by their odd appearance and unique behavior. While flying, they might easily be mistaken for the crane flies that occur in the same habitats, but they are not true flies at all. The most obvious difference is their possession of four wings rather than two. Instead, they are members of the so-called scorpionfly order (Mecoptera). When not flying among shaded tree lines near creeks and streams, they hang by their front legs from leaves and twigs and snatch other insects from midair with long legs that resemble grappling hooks. Sometimes they fly from one perch to another with prey still in their grasp. Courting males offer catches to females as "nuptial gifts" prior to mating. In the laboratory hangingflies have been known to eat as many as ninety small flies in less than two days (Setty 1931).

The Ottine swamps appear to be fine places to see and study these creatures. We encountered them in early June in a remote spot along Rutledge Creek, where they flew in such numbers that we easily captured

Fig. 9-8
Hangingfly (*Bittacus punctiger*) hanging from leaf

thirty specimens in the space of an hour. At the same time of year a few were seen in Palmetto State Park, caught in spider webs or hanging from the giant lettuce plants that were being defoliated by gangs of voracious swamp katydids. This particular species, the brown-spotted hangingfly, is believed to be one of the least common members of an uncommon group (Carpenter 1931; Setty 1940). The larvae resemble spiny caterpillars but feed as earthworms do, by eating soil (Byers 1987). In the eastern United States the adults appear in May and June rather than July and are classified as "early-season" (Sherman 1908).

Distribution: From the Atlantic Ocean to Texas, though there appear to be no records from quite a few of the intervening states. George Byers informed us that the University of Kansas collection has Texas specimens from Nacogdoches and Matagorda counties, both locations well east of the Ottine swamp. Perhaps our record marks the westernmost limit of the species as it is currently known.

Remarks: Length = 13 mm. Hangingflies are the only insects that capture their prey with their hind legs. The long, single-clawed appendages are used in this regard much as the front legs of a praying mantis are, but they are so strongly committed to their function that when hangingflies are confined in a jar with nothing to grasp, they form a tangled, writhing, helpless mass on the bottom of the container. They cannot use the legs for walking.

Similar species: Several hangingfly species might occur in the swamp, though we saw no others. There are only seven species in North America north of Mexico, and at the time of writing the only key to the species of the United States was already seventy years old (Carpenter 1931).

10
True Bugs

Generally speaking, to the public every insect is a bug. That is a fine use of a common name, but to an entomologist the word applies more strictly to all members of an insect group in the order Hemiptera, and to no other insects at all. Every true bug possesses mouthparts of the sucking rather than biting type, whether the bug is predatory and uses its beak to kill animal prey or whether it is parasitic and uses its beak to imbibe juices from living plant and animal hosts. All three feeding strategies may be observed while watching the true bugs of the Ottine swamps and marshes.

Cicadas

The cicada fauna of Texas is exceptionally diverse compared to that of other states on account of the state's size, its location at the crossroads of east and west, and its position within the Deep South so close to the Tropics. This fact was proclaimed long ago in a single-page journal article inviting others to study them here (Bromley 1933). At about that time a student at Texas A&M University, where Bromley spent considerable time, was doing precisely that, and on a comprehensive scale that promised to culminate in a much-needed identification key to the forty or more Texas species. But according to a legend passed down to us by Edward Riley, assistant curator of the Texas A&M University Insect Collection, the student became frustrated with the demands of his supervisor and brought the process to a halt by tossing the unfinished manuscript through an open window. A faculty member rushed outside to gather the scattered pages from the lawn, but the jaded student was gone forever. We examined a photocopy of this document "on location." No key had been constructed by the time of the episode, nor has one been published since. Thus, Bromley's invitation still stood seventy years later when these words were being written. It opens with the proclamation that "east central Texas is a cicada paradise," and should a second student finish what the first began, future attempts to identify the species of the Ottine swamps will encounter less difficulty than our own.

Fig. 10-1
Whistling cicada
(*Quesada gigas*)

Whistling cicada
Quesada gigas
(FIG. 10-1)

Biology: The queer song of the male was heard in and near Palmetto State Park, but we saw no individuals, despite the fact that this is the largest cicada in the wetlands. Nor did any come to our lights after dark, though they have a reputation for doing so.

The song is a startling whistle unlike the whirring chatter of other species. It has been compared to a steam whistle, a locomotive whistle, a popcorn whistle, and the whistle of a peanut roaster (Davis 1944). The latter two devices have passed from popular American culture so that many of us are no more familiar with them than with the insect. Our own comparison is to a teapot boiling on a stove. We were not even sure of the singer's identity as a cicada until we drew close enough to a tree to hear the brief, softer beginning of the song, which does contain a transient, more typically cicada-like chatter.

In Texas the whistling cicada has been reported from live oaks and mesquites, whereas in South America it is associated with palo blanco (*Vernonia patens;* Young 1980). Like the other cicadas treated here the juvenile stages known as nymphs or larvae live underground for a year or more, where they feed upon roots of various plants. Dry, brown husks left behind on tree trunks by emerging adults are seen more often than any living stage. The husks are empty juvenile exoskeletons.

Distribution: The whistling cicada has an enormous distribution from Argentina to Texas, which is apparently the only state in the United States where it occurs. We also heard it on a single occasion in the Lost Pines forest of Bastrop County. There it must be within a few short miles of its northern geographical limit. One of us lived for decades in nearby College Station but never heard the whistler's song.

Remarks: Length = 50 mm. Other common names include locomotive cicada and soup bug, the latter name derived from its habit of flying to domestic lights and landing among the dishes (Davis 1944). Its size and scientific name also warrant the appellation of giant cicada. The specimens shown here were collected in La Union, Guatemala, in March 1992. We thank Lawrence Forcella for providing them.

Similar species: None. The male's song is more unmistakable than its striped appearance, and according to our own experience a sighting is unlikely in any event.

Superb cicada
Tibicen superba
(FIG. 10-2)

Biology: We encountered the superb cicada in Palmetto State Park, often near the stone water tower and always after dark when the vulnerable adults were exiting their old exoskeletons on stones, trees, and herbaceous vegetation. We found no information regarding food plants or the song of the male.

Distribution: From the Mississippi Valley region to a western limit in New Mexico.

Remarks: Length = 29 mm. The superb cicada is the common urban and suburban species of Austin, College Station, and other central Texas cities. Its song is nearly synonymous with the region's long, hot summers.

Similar species: None. The mostly green thorax with its four strongly contrasting, black, bullet-shaped markings is distinctive.

Fig. 10-2
Superb cicada (*Tibicen superba;* left)
and margined cicada
(*Tibicen marginalis;* right)

Margined cicada
Tibicen marginalis
(FIG. 10-2)

Biology: The margined cicada flew to our lights stationed after dark just above the banks of the San Marcos River. Males sing with a sound described as "Z'we, Z'we" (Beamer 1928) and may be heard from morning into the hours after midnight. Mated females lay their eggs in living or dead limbs of willow, maple, and cottonwood (Beamer 1925, 1928). Thus, the species is characterized as a floodplain denizen.

Distribution: From the Atlantic Ocean to the Rocky Mountains.

Remarks: Length = 37 mm. Male choruses in south-central Texas can be so loud that "one cannot carry on a conversation in the vicinity" (Davis 1935, 176).

Similar species: The two white dashes on the back are not the margins referred to in the animal's name. Other cicadas occurring in the area but treated elsewhere are the resh cicada and the hieroglyphic cicada (Taber and Fleenor 2003b).

Waxy cicada
Tibicen pruinosa
(FIG. 10-3)

Biology: We found this cicada transforming to the adult stage on vegetation in Palmetto State Park during the summer months. Its song is a "z-zape,

Fig. 10-3
Waxy cicada (*Tibicen pruinosa*) with shed skin of underground juvenile stage

z-zape, z-zape" or "za-wie, za-wie, za-wie" that is said to be unique to the species (Davis 1910; Beamer 1928), yet others have described it as a repeated "zip" punctuated by "twang-twang" (Barber 1910). The song, however one conceives it, often commences in late afternoon and continues until dark. Individuals also come to lights after dark, and sometimes they cause a commotion when they fly into places of business (Froeschner 1952).

Females lay eggs in living or dead tree tissues, and hatching juveniles tumble to the ground below, where they dig into the soil to begin feeding on roots, as is the case for the other cicadas treated here. Preferred trees in these wetlands are ash, elm, and maple (Beamer 1928).

Enemies wait at every turn. By night cats and toads hunt the vulnerable, flightless juveniles when they emerge from underground (Davis 1922). During daylight hours the cicada-killer wasp is a potent enemy that stings the adult, drags it underground, and lays an egg on the helpless creature. A grub hatches from the egg and eats the paralyzed but living cicada.

Distribution: From the Atlantic Ocean to Texas.

Remarks: Length = 27 mm. Common names applied to this and to other cicadas that spend one or only a few years underground are dog-day cicada and harvestman. A tendency to sing at night has also earned this one the epithet of evening cicada.

Similar species: Of those species treated here the waxy cicada may be identified by its photograph. Similar species might occur in the wetlands, but no identification key was available at the time of this writing.

Fig. 10-4
Giant electric light bug (*Lethocerus medius*)

Giant electric light bug
Lethocerus medius
(FIG. 10-4)

Biology: This formidable aquatic insect is a powerful, air-breathing predator that eats other insects, small fishes, salamanders, tadpoles, and even snakes. It is the largest true bug in the swamps, and its ambush strategy is aided by such great resemblance to a dead leaf that its unsuspecting prey sometimes seeks shelter beneath the bug's body.

  Female giant electric light bugs lay their eggs above the waterline on emergent vegetation. The male crawls out of the pond to guard them, and when approached by a human, it brazenly spreads the powerful front legs and postures in defiance (Smith and Larsen 1993).

Distribution: Within the United States this species occurs from west-central Arizona to an eastern limit near Houston, Texas, not far from the Ottine wetlands. Thus, it is one of the relatively few western animals in the swamps.

Remarks: Length = 60 mm. Electric light bugs can inflict a painful stab if handled. The common name reflects a habit of flying in large numbers to streetlights.

Similar species: Several giant electric light bug species might occur in the swamps, but every individual we identified was *L. medius.* The species do resemble one another closely, and keys must be consulted for a confident identification.

Fig. 10-5
Toe-biter (unidentified *Belostoma* species) perched atop stem in cordgrass marsh, perhaps escaping the water snakes that were foraging below

Toe-biters
Belostoma confusum, B. fusciventre, B. lutarium
(FIG. 10-5)

Biology: Toe-biters do indeed appear to be miniature versions of their much larger relative, the electric light bug. One difference is the manner of guarding the eggs, which is perhaps unique in the animal kingdom. The female deposits her clutch on the male's back, sometimes in the face of resistance from her mate. This has two potential benefits. First, the eggs are aerated more efficiently when moved about through the water with the swimming adult. Second, they are protected at least to the degree that the adult protects itself. This leads one to wonder why the behavior is not more widely adopted in the animal kingdom.

Distribution: The confused toe-biter (*B. confusum*) occurs only in southern Texas and in southeastern Arizona, whereas the spotted species (*B. fusciventre*) is limited to central and southern Texas and Louisiana. The yellow toe-biter (*B. lutarium*) has a broader range than either, from the Atlantic Ocean to central Texas.

Remarks: Lengths =17 mm to 25 mm. The common name of these aquatic predators reflects their habit of biting the toes of people who venture barefoot into ponds.

Similar species: Additional species might occur in Ottine ponds and lagoons. Confident identification requires microscopic examination and the proper keys. Yet as a group they do not resemble any other bugs save the much larger electric light bug.

Southern water scorpion
Ranatra australis
(FIG. 10-6)

Biology: This lanky, aquatic predator resembles a praying mantis in shape, size, means of capturing its prey, and in its diet of other insects; but it differs from the land dweller by the inclusion of fishes and tadpoles among its prey. The tail-like structure responsible for the water scorpion's common name is actually a harmless tube used for breathing air when the rest of the bug hangs head down just beneath the surface. The long, dark, slender body and a tendency to feign death when disturbed provide excellent camouflage at the bottom of a dip-net dripping with muddy leaves and twigs dredged from the bottom of a pond.

    This is not a swimming bug but one that gets around underwater by walking. On land it is very clumsy. In the air it is a good flier. Flight is seldom witnessed and might only occur when drying ponds require dispersal to new waters. Females lay their distinctive eggs in a wet log. The eggs have filaments that aid in the acquisition of oxygen for the young bug developing inside.

Distribution: From the Atlantic Ocean to central Texas.

Fig. 10-6
Southern water scorpion (*Ranatra australis*) with parasitic mites

Remarks: Length = 65 mm. Water scorpions cannot sting, but they can deliver a painful stab with the piercing mouthparts. They also discourage capture by rubbing the front legs against the body to make squeaking noises.

Under the microscope this submariner's body is often seen to be covered with algae, parasitic mites, and sundry debris that evoke an image of seaweed and barnacles growing on a ship's hull. More exaggerated names than water scorpion are alligator flea and water dog.

Similar species: Four or five slender water scorpion species of similar size and appearance might creep about in these still waters. Reliable identification requires the proper keys (Sites and Polhemus 1994). The remaining species are the four-toothed water scorpion (*R. quadridentata*), the dark water scorpion (*R. nigra*), Bueno's water scorpion (*R. buenoi*), and the Texas water scorpion (*R. texana*). It is disappointing that the one species we never saw is the only water scorpion endemic to central and southern Texas and is also named for the state (Sites and Polhemus 1994). The dark water scorpion and Bueno's water scorpion are eastern species that occur from the Atlantic Ocean to a western extreme in or near the wetlands of central Texas. The four-toothed species is unusual for being a western species. It ranges from California to east Texas.

Howard's water scorpion
Curicta scorpio
(FIG. 10-7)

Biology: The biology of Howard's water scorpion is similar to that of the larger, more elongate, and more tubular water scorpions.

Distribution: Found nowhere else in the United States but in southern Texas and southern Louisiana.

Remarks: Length = 26 mm. Howard's water scorpion is generally considered to be a rare or uncommon species. It might be locally abundant in the ponds and lagoons of the Ottine swamps.

Similar species: None.

Kissing bugs
Triatoma sanguisuga, T. indictiva
(FIG. 10-8)

Biology: Adult kissing bugs stab mammals, including humans, with sharp, tubular beaks and suck blood from the host's body. Thus, their diet is identical to that of the mosquito, and it is their supposed habit of biting

Fig. 10-7
Howard's water scorpion (*Curicta scorpio*)

Fig. 10-8
Black kissing bug (*Triatoma indictiva;* left) and red and black kissing bug (*Triatoma sanguisuga;* right)

humans about the mouth while they sleep that gave coin to the common name of kissing bug. Strangely enough, reports differ markedly as to the severity of the stab or "bite." According to some the attack is painful (Readio 1927 and sources cited therein), but according to others it is painless (Lent and Wygodzinksy 1979). Perhaps both are correct. Perhaps it behooves the bug to feed without disturbing the host, whereas in self-defense it is obviously more adaptive to cause pain. Either way, the red and black kissing bug (*T. sanguisuga*), the more common of these two "bloodsucking cone-noses," is an established vector of the neurological sleeping sickness known as Chagas' disease. The causative agent of Chagas' disease is a protozoan, but it is not transmitted by the bite itself. The disease agent is transmitted via the excrement of the bug if and when this is accidentally rubbed into the wound, into the eyes, or into the mouth after the blood meal is completed.

Particularly relevant to the Ottine swamp is the association between kissing bugs and palmetto fronds, where they coexist elsewhere and perhaps here with green tree frogs (*Hyla cinerea*). It is not known if the bug actually feeds on the amphibian, but it is known that opossums and armadillos, both abundant in the swamp, are important natural reservoirs of Chagas' disease. We did not find kissing bugs in any habitat that might be described as natural, not even beneath logs or stones. Instead we found both species after dark at an uninhabited stone water tower and the red and black kissing bug also at black lights. Perhaps they feed in the tower upon the abundant introduced smokybrown cockroaches (*Periplaneta fuliginosa*). They are certainly known to feed upon roaches elsewhere. Or perhaps they feed upon small mammals living in the building.

It should be remembered that the biological details of the previous summary apply to the red and black species. Little is known about the much rarer black kissing bug (*T. indictiva*) except that it had not been reported as a vector of Chagas' disease at the time of this writing. We captured a single specimen.

Distribution: The red and black kissing bug occurs from the Atlantic Ocean to a western limit in Arizona, whereas the black species is one of the few western animals we saw in the swamp. It occurs from Arizona to an eastern limit near Dallas, Texas.

Remarks: Length of *T. sanguisuga* = 21 mm; length of *T. indictiva* = 27 mm. Other common names for *T. sanguisuga* include Texas bedbug, Mexican bedbug, big bedbug (the animal does bite sleeping humans at night), bellows bug, and, our favorite, Arizona tiger. *Triatoma indictiva* is too poorly known to have a common name except for the one given here on the basis of its black color.

Similar species: At least seven kissing bug species occur in Texas, and it is wise to recognize their general appearance and to avoid them. Confident identification to the species level requires the use of formal keys. At the time of writing, the authoritative work was more than twenty years old (Lent and Wygodzinksy 1979).

Fig. 10-9
Wheel bug (*Arilus cristatus*) feeding on female giant walking-stick

Wheel bug
Arilus cristatus
(FIG. 10-9)

Biology: This animal is a predator that kills its prey, unlike kissing bugs that merely take a small drink of blood. It is the largest assassin bug in the wetlands, and there are few terrestrial or flying insects that it cannot capture and kill. Wheel bugs perch on leaves and lurk in rank vegetation along the shores of the oxbow lake.

Distribution: From the Atlantic Ocean to New Mexico.

Remarks: The wheel bug is named for the remarkable cogwheel structure on its back. Its function is unknown, and because both sexes bear wheels of similar proportions, it probably has not evolved to its current large size by female choice or by combat between competing males. This insect can deliver a painful stab if handled, with effects lasting for weeks. Considering the wheel bug's size and power, it is perplexing to discover that its color is dull gray rather than the bright red and black of warning.

Similar species: None.

Corsair
Rasahus hamatus
(FIG. 10-10)

Biology: Corsairs are colorful, nocturnally active assassin bugs that fly to black lights at night and are seldom seen under other circumstances. Like

Fig. 10-10
Corsair (*Rasahus hamatus*) at black light

most assassin bugs they are predatory in habit. Its previously unreported diet includes the small cricket species that one specimen was eating at our blacklight trap.

Distribution: This is a tropical species ranging as far north as the southern United States. The distribution is remarkably disjunct, with separate populations occurring in Texas, Florida, Oklahoma, and Missouri.

Remarks: Length = 20 mm. The corsair is said to occur under rocks and boards and in vegetation along wet lowlands. We never saw the big, attractive bug in such places, and we never saw it during the day. It delivers a painful bite when handled.

Similar species: All three of the corsair's closest U.S. relatives occur in Texas, and confident identification requires a microscope and the proper keys.

Mole cricket assassin bug
Sirthenea stria
(FIG. 10-11)

Biology: We encountered one specimen in Palmetto State Park and none in the nearby Lost Pines forest, though we spent years looking for such species. The very attractive insect lives underground in the tunnels of its prey, the southern mole cricket, sometimes flying to lights at night (Hudson 1987; Willemse 1985). Juveniles are generalist predators, but the adults become such specialized predators of mole crickets that they will

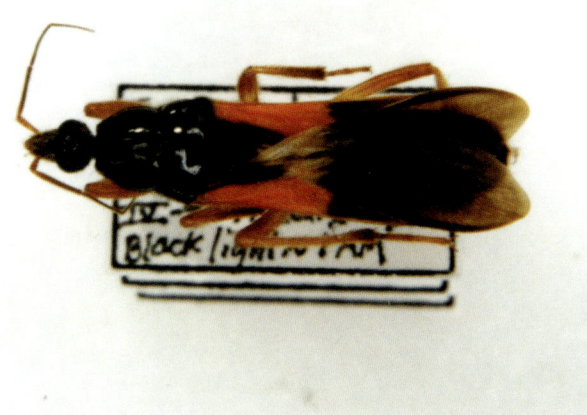

Fig. 10-11
Mole cricket assassin bug (*Sirthenea stria*)

apparently starve rather than feed upon other cricket species that are offered to them.

Distribution: In the northern part of its range the animal apparently occurs from the Atlantic Ocean to a western limit in Texas (Willemse 1985), though there is some disagreement on this, and representation west of Texas is likely to be spotty at best.

Remarks: Length = 25 mm. The rarity of this assassin bug in central Texas is remarkable given its broad geographic range within the New World, and perhaps this is explained by the relative rarity of its prey. It is better known by the scientific name *S. carinata*.

Similar species: The pinkish orange of the wings is unique, and there is no other species of this genus in the United States.

Black May beetle-eater
Melanolestes abdominalis
(FIGS. 10-12–10-13)

Biology: This black and red assassin bug lurks beneath rocks in the daytime and preys upon May beetle adults and their grubs (Readio 1927). When attacking adults, the bug mounts the beetle from behind and pierces the neck with its beak. Spongy pads on the predator's legs stick to the prey much like suction cups.

Males have fully developed wings, but adult females appear to be juveniles at first sight because their wings are tiny and probably useless. The

Fig. 10-12
Black May beetle-eater (*Melanolestes abdominalis*) exposed from beneath rock

Fig. 10-13
May beetle feeding on foliage; potential prey of the black May beetle-eater

mating process is very similar to the capture of prey. Males mount females from behind and place the beak in the same neck area that is pierced when killing May beetles. The female makes scraping sounds during copulation by rubbing her beak against her body. When mating is concluded, the female lays eggs one at a time by pushing them into the soil beneath a rock. This species flies to black lights in spring.

Distribution: From coast to coast within the United States.

Remarks: Length = 20 mm. The bite of the black May beetle-eater is painful, and when it strikes a human face while flying, it is apt to bite immediately (Blatchley 1926).

Similar species: Some authors recognize *M. abdominalis* (with red color on the abdomen) and *M. picipes* (which is entirely black) as two different species. Others believe that there is only one species that varies in color. Here we consider them as two different species and assume that the biology of both is essentially identical. Red and black individuals are more common than those that are entirely black. In the wetlands we know of no other assassin bug similar enough to cause confusion.

Red and black assassin bug
Zelus longipes
(FIG. 10-14)

Biology: The specimen pictured here was feeding on a small fly at the oxbow lake. Caterpillars are an important part of the diet, and therefore the bug is considered beneficial.

Distribution: From Canada to South America.

Remarks: Length = 12 mm. This brilliant, slender species bears the red and black warning colors so common among insects that bite and sting.

Similar species: None.

Fig. 10-14
Red and black assassin bug (*Zelus longipes*) with prey on poison ivy leaf, and Louisiana deervetch flower (*Vicia ludoviciana;* right)

Spine-headed bugs
Acanthocephala declivis, A. confraternus
(FIGS. 10-15–10-16)

Biology: The giant spine-headed bug (*A. declivis*) is the largest terrestrial bug we saw. It probably sucks sap from oaks and other trees, but through the tough bark rather than through the tender surface of a leaf. We never saw it in the act, though we often saw the smaller common spine-headed bug (*A. confraternus*) sucking sap through the bark of green ash. Individuals of

Fig. 10-15
Giant spine-headed bug (*Acanthocephala declivis*)

Fig. 10-16
Mating pair of the common spine-headed bug (*A. confraternus*); each individual has a white egg of a parasitic insect on its back.

this species often have a small white egg attached to their back, presumably the egg of a parasitic fly. We are unaware of previous reports of food plants for either insect. A dozen or so of the larger species were found overwintering in a clump of cactus, on the lower surface of the pads where they were out of sight while gaining protection from the formidable spines.

Spine-headed bugs possess curious hind legs that bear spines and flat dilations. The shapes of these structures have earned these bugs the alternative name of leaf-footed bug. Individuals of both sexes use their specialized legs to seize and pinch one another in battles over food and reproduction (Eberhard 1998).

Distribution: The giant spine-headed bug occurs from the Atlantic Ocean to Arizona and south into the Tropics. The common spine-headed bug occurs from the Atlantic Ocean to Texas.

Remarks: Length = 30 mm. Of land-dwelling true bugs the length, but perhaps not the bulk, of the giant species is rivaled only by that of the predatory wheel bug. Its relative has a length of 20 mm. Both are much smaller than the aquatic electric light bug. The resemblance to the assassin bug extends even to shape and color, but this is a sap feeder that will not bite if handled. Instead, it releases a pleasant-smelling chemical that melted our plastic collecting containers.

Similar species: At least one additional spine-headed bug occurs here, though its occurrence is uncommon compared to its occurrence in central Texas outside the wetlands. Males of this species, aptly named *A. femoratus* (length = 28 mm), have swollen, bowed hind legs with strong spines and are easily recognized. The female is not so readily distinguished. A fourth and fifth species might also occur in the wetlands. These are *A. terminalis* and *A. thomasii,* but the disappointing and nearly century-old key to the species of this genus lacked simple line drawings that would have made identification of these bugs an easier task (Gibson and Holdridge 1918).

Predaceous stink bug
Apoecilus cynicus
(FIG. 10-17)

Biology: This insect resembles the more familiar plant-feeding stink bugs, but it approaches 25 mm in length and is a voracious predator of caterpillars. We saw many brightly colored nymphs of various sizes feeding on prey larger than themselves in Palmetto State Park. They were perched on leaves of trees and shrubs with their beaks pointing straight out in front

Fig. 10-17
Predaceous stink bug
(*Apoecilus cynicus*)

with the impaled caterpillar at the tip. One prey species is the hackberry butterfly (*Asterocampa celtis*). We saw a juvenile bug patrolling the large, protective, silken web covering dozens of webworms. From time to time the would-be predator probed the web with its beak in its efforts to spear a caterpillar inside.

Distribution: From the Atlantic Ocean to Arizona.

Remarks: Length = 20 mm. Several sources describe this species as uncommon. In the Ottine wetlands in April it is briefly abundant as both adult and juvenile, and not surprisingly so, as caterpillars are abundant in these swamps and marshes. We confirm the reported preference for tree line habitats at the margins between woods and clearings.

Similar species: None that we saw.

Green stink bug
Acrosternum hilare
(FIG. 10-18)

Biology: Green stink bugs fly to black lights in spring and are more often seen there than on their many host plants. Hosts occurring in the swamp and adjacent uplands include grape, elderberry, hackberry, oak, ironweed, hawthorn, sumac, ash, pecan, peppervine, boxelder, mulberry, buttonbush, trumpet-vine, and Virginia creeper (McPherson 1982).

Distribution: From coast to coast within the United States.

Remarks: Length = 19 mm. This fruit-feeding bug defends itself with a mal-

Fig. 10-18
Green stink bug
(*Acrosternum hilare*)
perched on sassafras
twig

odorous chemical that ruins the taste of hand-picked dewberries and mulberries after it has sucked its fill of the fruit's juices.

Similar species: The pestiferous southern green stink bug (*Nezara viridula*) is a similar and much better known species. Identification relies upon microscopic examination with the aid of a formal key.

Ash stink bug
Menecles insertus
(FIG. 10-19)

Biology: This gregarious, nocturnal animal has a remarkable daily schedule that includes migration up and down the trunk of a tree. It hides on the ground beneath leaf litter during the daylight hours, and when darkness falls, the bark-colored creature begins crawling up trunks in large numbers. After feeding on the trunk throughout the night, it descends once more into the fallen leaves where its brown, mottled coloration continues to provide camouflage against potential predators (Balduf 1945). Known hosts occurring in the wetlands include elm, hickory, and hackberry. In the Ottine swamps green ash is a preferred host, but this appears to be the first record of the insect from the tree. We often saw several to many bugs feeding in bark crevices after dark in Palmetto State Park.

Distribution: The ash stink bug has one of the most unusual distributions of the insects in the wetlands. Though occurring coast to coast within the United States, it is entirely absent from the southeast with the exception

Fig. 10-19
Ash stink bug
(*Menecles insertus*);
mating pair

of perhaps a single record from Florida. A search of the literature, including a Web search with the OVID service, suggests that our record from the Ottine swamps and from the Lost Pines are the first reports of this species from Texas.

Remarks: Length = 14 mm. This bug is widely distributed "but nowhere abundant" (Van Duzee 1904) and is not common enough to earn a place in one of the standard works on the true bugs of the United States (Slater and Baranowski 1978). It is curious that we never found the animal on the very abundant hackberry trees. When these bugs are feeding in numbers on ash, the sight suggests huge ticks parasitizing a botanical host. This is true of the juveniles also, though their habits seem not to have been published before despite interest in the diet of juvenile stages (Esselbaugh 1948).

Similar species: Only two species of this genus occur in the United States, and there is little chance for confusion because the ash stink bug's close relative is the Texas endemic *M. portacrus,* which is restricted to the Big Bend country hundreds of miles west of the swamps. The brown stink bug (*Euschistus servus*) is a crop pest that occurs from coast to coast. With a hand lens the margin of the body just behind and on either side of the head can be seen to display an array of tiny teeth. If these are present on the ash stink bug, they are not obvious at such low magnification.

Fig. 10-20
Soapberry bugs
(*Jadera
haematoloma*)
aggregating and
mating on a log

Soapberry bug
Jadera haematoloma
(FIG. 10-20)

Biology: This is a striking blackish gray bug with bloodred shoulder patches. Its color advises potential predators of its unpalatability, which is due to its habit of feeding on seeds of members of the soapberry family. Native host plants in Texas include the soapberry tree (*Sapindus saponaria*) and the heartseed vine (*Cardiospermum halicacabum*). We found large aggregations containing dozens of individuals on logs near Rutledge Creek in midwinter in areas where balloon-vine was abundant during the previous growing season. These groups contained nymphs and mating adults, with some individuals bearing short wings, others with long wings. Large groups might help to strengthen the effect of red warning coloration on potential predators.

    Males prevent other males from mating with their own mates by remaining attached to the female until she lays eggs, a time interval lasting as long as eleven days (Carroll 1991). Eggs may be deposited in soil or in tissues of the host plant itself.

Distribution: The soapberry bug occurs spottily from coast to coast with greatest density in the southern plains states, especially Texas.

Remarks: Length = 14 mm. In Texas the beak of the soapberry bug is said to have evolved to greater lengths in only a few decades after the introduction of a new host plant, the heartseed vine (*Cardiospermum halica-*

Fig. 10-21
Louisiana hibiscus
bug (*Niesthrea
louisianica*)

cabum; Carroll and Boyd 1992; Carroll and Dingle 1996), but as far as we
know, the plant is native.

Similar species: Two additional species of this genus occur in Texas. Neither
has the dark body with red lateral stripes. We thank Scott Carroll for help
with the identification of this animal.

Louisiana hibiscus bug
Niesthrea louisianica
(FIG. 10-21)

Biology: The hibiscus bug is a small, colorful, parasitic insect clad in yellow
and red and flecked with small dark spots. We found a congregation feeding upon halberd-leaved hibiscus (*Hibiscus laevis*) in North Soefje Marsh.
It also occurs on other members of the same plant family, such as okra,
velvetleaf, and rose of Sharon (Sailer 1961), sucking sap from seed pods
and flower buds.

Distribution: From the Atlantic Ocean to Arizona.

Remarks: Length = 8.5 mm. This New World parasite finds the introduced
Old World velvetleaf more suitable as a host than its natural or native
plants (Jones et al. 1985).

Similar species: Texas is the only state in the United States where all three
species of hibiscus bug occur. A formal key is required to distinguish one
from another (Chopra 1973), though the Louisiana bug is by far the most
commonly encountered.

11

Crustaceans, Millipedes, and Centipedes

The animals featured in this chapter and all those that follow are invertebrates, but none of them are insects. This is evident from the fact that all adult crustaceans, millipedes, and centipedes have more than six legs. Some evolutionists believe that of the three, insects are most closely related to the many-legged millipedes and centipedes. Others believe that insects are more closely related to the crustaceans, which, like the insects, bear relatively few pairs of legs. Because crustaceans are more closely tied to water, we expect their diversity in the Ottine wetlands to be greater than the diversity of the fully terrestrial millipedes and centipedes.

Gulf Coast crawfish
Procambarus zonangulus
(FIG. 11-1)

Biology: Crawfishes are gilled aquatic crustaceans that resemble small lobsters. Their diet is omnivorous, and although plant matter is one of the preferred foods, they do capture and kill animals. We netted them in ponds during the day and picked them up from dry ground when they wandered about after dark.

In areas where water is ephemeral the animals must go underground to survive the drought of summer. Burrows are marked by tall mud chimneys up to eight inches in height. These are constructed from mudballs carried up from below, but their function is uncertain. They sometimes occur in numbers large enough to suggest a subterranean city.

Distribution: Gulf Coast crawfishes occur from the Atlantic Ocean to an undetermined western limit in Texas (perhaps the Ottine wetlands itself;

Fig. 11-1
Gulf Coast crawfish
(*Procambarus zonangulus*).

Huner 2002). It appears that this animal, usually reported as *P. acutus*, has never been reported from these wetlands, from Gonzales County, or from the Guadalupe and San Marcos river systems.

Remarks: Length = 100 mm. The Gulf Coast crawfish is a cultivated culinary species as is the commercially better-known red swamp crawfish (*P. clarkii*), which we did not see in the swamps.

Similar species: Crawfishes are difficult to identify, and additional species undoubtedly occur in the Ottine wetlands. Formal keys require not only the male sex, but a particular *type* of male known as "form I." These are individuals in breeding condition (Penn and Hobbs 1958; Hobbs 1989; Hobbs and Hobbs 1990).

Clam shrimp
Lynceus gracilicornis
(FIG. 11-2)

Biology: Clam shrimps are neither, though, like true shrimps, they are aquatic crustaceans and, like true clams, they are enclosed in a bivalve shell. We encountered the tiny, bizarre creatures when they swam into underwater light traps submerged in the shallow water of ephemeral lagoons in Palmetto State Park.

This species is known to prefer pond margins and is most abundant in spring. Individuals swim with the aid of antennae as well as legs, often upside down or even in a spiral motion. Their food is plankton gathered from the water as they move through it. Males use hook-shaped legs to

Fig. 11-2
Clam shrimp
(*Lynceus gracilicornis*) from lagoon in
wetland 1

grasp their mates, and females carry up to two hundred eggs (Martin, Felgenhauer, and Abele 1986; Martin and Belk 1988).

Distribution: At least patchily distributed from the Atlantic Ocean to central Texas.

Remarks: Length of shell = 6 mm. Down to its ducklike face this odd creature is reminiscent of the duck-billed platypus because it seems to be constructed of parts taken from very different animals. In life it is orange to rose in color, with a maroon shell.

Similar species: Other clam shrimps might occur in the Ottine swamps. A microscope and formal keys are necessary for confident identification.

Water-hog louse
Caecidotea obtusa
(FIG. 11-3)

Biology: The water-hog louse is an aquatic crustacean of stagnant rather than running waters. It is related to the more familiar terrestrial pillbugs or sowbugs that are so common beneath rocks and logs in the wetlands. We were alerted to the presence of the aquatic species only when we placed incandescent lights in submerged water traps and lowered them into the lagoons of Palmetto State Park during the spring of the year. They swam into the traps toward the bright light source but not in large numbers. We saw no other report indicating capture by this means.

Little is known about the biology of the water-hog louse occurring in the Ottine wetlands, and in a South Carolina study it was described as

Fig. 11-3
Water-hog louse (*Caecidotea obtusa*) from lagoon in wetland 1; in spoon for scale

very rare (Biernbaum 1989). If its close relatives provide an accurate guide, then it is a bottom-sprawling animal that feeds on decaying plant matter. Males are remarkable for their large mantislike forelimbs, which are used to seize females (Ellis 1961) and perhaps to do battle with other males. The clawlike tips of these legs are also useful for the identification of species. Females use their legs to brood the young in pouchlike structures. Juveniles consume their own shed skins before maturation.

The breeding period is likely to be spring and perhaps early summer, before the largely ephemeral lagoons dry up completely in the summer heat. A related species of the eastern United States lives in alkaline water with a pH of 8.4 and with considerable organic material. In fact, it seemingly does best in at least moderately polluted water (Ellis 1961). Water-hog lice might not occur in the more acidic peatlands of the Ottine area. When we tested the lagoons in winter, the water had a neutral pH of 7.0.

Distribution: This appears to be the first record of the species from Texas. It was previously known to occur only within a narrow coastal band from Florida to Louisiana (Williams 1970; Biernbaum 1989).

Remarks: Length = 10 mm. In lieu of a suitable American common name we have chosen the name used in Britain for a related species. The scientific name has changed radically since it was made known to science in 1970, and its former name of *Asellus obtusus* remained a more familiar one at the time of this writing.

Similar species: We found no other aquatic isopod in our submerged light traps, and we know of no other surface-dwelling water-hog louse reported from Texas. Others might coexist in the lagoons of Palmetto State Park, and a formal key is required for confident identification. A few species might live in subterranean waters. If present at all, these are of course unlikely to be encountered.

Fig. 11-4
Common pillbug
(*Armadillo vulgaris*)

Common pillbug
Armadillo vulgaris
(FIG. 11-4)

Biology: This terrestrial crustacean, noted in both its common and scientific name for a defensive habit of rolling up into a ball when disturbed, is abundant after spring rains, when it can be found by the dozens beneath logs and stones, often in the company of snails.

Its omnivorous diet includes the remains of plants and animals. In Texas common pillbugs are also pests of living cotton; in Louisiana, pests of cucumbers; and in Virginia, pests of cultivated mushrooms (Richardson 1905).

Pillbugs die off in great numbers during drought and during floods. Two plants are known to provide havens or "refugia" in these times. If the Ottine swamps are comparable to conditions in California, then the animals avoid desiccation by moving into the moist bases of Roosevelt weed (*Baccharis neglecta;* Paris 1963). Drowning is averted by aggregating within clumps of Gulf cordgrass (*Spartina spartinae;* Miller and Cameron 1987). One living natural enemy is the slender ponerine ant (*Leptogenys elongata*), which has a diet that specializes in pillbugs (Wheeler 1904).

Distribution: This is an Old World species repeatedly introduced into the United States from the Atlantic Ocean to the Pacific. Like the red imported fire ant it invades new territory with nursery stock such as trees, shrubs, and other ornamentals.

Remarks: Length = 18 mm. Females often display a mottled coloration not visible on males. This is caused by the accumulation of a chemical substance. The species is better known by its alternative name of *Armadillidium vulgare,* though this was discarded shortly before our studies began (Opinion 1897).

Similar species: A close relative might occur in the swamp. It can be distinguished by a notch on the front of the head, which *Armadillo vulgaris* lacks (Muchmore 1990). Presumably this species, known at the time of writing as *Armadillidium nasatum,* will also be renamed as *Armadillo nasatum.*

Waxy woodlouse
Metoponorthus pruinosus
(FIG. 11-5)

Biology: The waxy woodlouse lives in large numbers among moist grass roots and wet rocks and in urban areas within the cover of greenhouses (Collinge 1946). In the Ottine swamps we found it hiding beneath rocks and decomposing logs. The function of the bluish gray waxy, or "pruinose," coating, when present, is not agreed upon. Some believe it reduces the chance of death by desiccation, but this has been disputed (Collinge 1946).

Fig. 11-5
Waxy woodlouse
(*Metoponorthus*
pruinosus)

Distribution: Like the common pillbug this crustacean is an introduced species, probably of European origin, that now occurs coast to coast within the United States. It is the most widely distributed terrestrial isopod in the world (Garthwaite and Sassaman 1985).

Remarks: Length = 11 mm. Waxy woodlice are flatter and faster than the common pillbug. Perhaps for the latter reason they are not as apparent during a log-rolling search for land-dwelling crustaceans. It is notable that both animals, the most abundant terrestrial isopods even in wild, swampy areas, are introduced species that are not native to Texas.

Similar species: The waxy woodlouse was considered to be a single widespread species until its total population within the United States was judged to consist of at least two species (Garthwaite and Sassaman 1985). According to the map provided therein, any specimens collected as far south as Texas are expected to be individuals of their newly recognized species, *M. floria* (described in the paper as *Porcellionides floria*). Intriguingly, though the Ottine swamps are located well within the southern half of this southern state, the specimens do appear to be individuals of the long-known and more characteristically northern species *M. pruinosus*. Laboratory experiments suggest that the northern species will replace the southern species whenever the two are brought into competition in the wild (Garthwaite and Sassaman 1985). Perhaps this is happening in the Ottine swamps.

  Metoponorthus virgatus is a third species that might occur here. It is known from sandy soil in southern Texas, and in Florida, from Palmetto scrub (Van Name 1936).

Giant millipede
Narceus americanus
(FIG. 11-6)

Biology: This is an attractive reddish brown animal that may exceed 153 mm in length. It is uncommon in the wetlands and is more at home in forested uplands nearby. Wherever the millipede occurs, it usually remains hidden beneath logs during the day. On warm nights it emerges to crawl on the ground and on tree trunks in search of the dead plants and animals that it eats.

  Molting occurs during dry periods when the animal burrows into a log, seals itself off, and sheds its skin (Hopkin and Read 1992). Its eggs are laid in a curious manner. The female places each one in a wad of chewed leaf litter, passes it backward with many cooperating legs, and shapes it

Fig. 11-6
Giant millipede (*Narceus americanus*)

Fig. 11-7
Centipede (probably *Hemiscolopendra marginata*) foraging at night five feet aboveground on tree trunk

within the rectum before placing the egg in a pile with others (Levi and Levi 1987).

Distribution: From the Atlantic Ocean to a western limit near Georgetown, Texas, not far north of the Ottine swamps.

Remarks: Length = 150 mm. Giant millipedes do not bite or sting when handled. They do dispense a yellow fluid that stains the skin but does no real harm.

Similar species: None. Several smaller millipede species of attractive orange or red colors are occasionally seen here, usually hiding beneath rocks and logs but sometimes crawling on bark in the manner of the giant American

Fig. 11-8
An orange millipede
(*Eurymerodesmus*
undetermined
species)

millipede. Millipedes can be difficult to identify, and the experts themselves sometimes cannot provide identification even when material is sent for examination. Centipedes are exclusively predatory and have only one pair of legs on most segments rather than two (Fig. 11-7).

Orange millipedes
Eurymerodesmus dactylocyphus, E. amplus
(FIG. 11-8)

Biology: These are beautiful orange and brown or orange and olive millipedes reaching lengths of just under 25 mm. We found *E. dactylocyphus* beneath logs in Rutledge Swamp and in Palmetto State Park. The few specimens previously collected were found among sedge-grass roots, on a shady river bank, and near a fence (Shelley 1989). *Eurymerodesmus amplus* is found under similar cover but was also seen at night with the aid of an ultraviolet light, presumably fluorescing in the glow of the lamp. There are no reports of diet or behavior in either species. Millipedes in general are scavengers of decaying plant material.

Distribution: *Eurymerodesmus dactylocyphus* is a Texas endemic that occurs only in a narrow corridor between Victoria in the south and the Waco area in the north. *Eurymerodesmus amplus* has a much wider range though still restricted compared to that of most animals occurring in the swamp. It ranges from Louisiana to central Texas. These are the first records of this millipede genus from Gonzales County.

Remarks: Length = 25 mm. The orange and brown color of *E. dactylocyphus* was unknown until now because only preserved specimens were available when the species was made known to science.

Similar species: Millipedes are difficult to identify because male individuals, a microscope, and formal keys are usually necessary. We thank Rowland M. Shelley for helping us identify our specimens. Several additional species probably occur in these wetlands.

12
Spiders and Scorpions

All invertebrates treated in this chapter are arachnids rather than insects. Thus, they possess eight legs rather than six. The dangerous, bark-dwelling brown recluse spider (*Loxosceles reclusa*), so abundant beneath tree bark in the nearby Lost Pines forest, is seldom encountered here. The same is true to a lesser extent for the potentially deadly southern black widow (*Latrodectus mactans*). Both are treated in a companion volume dedicated to invertebrates more characteristic of relict upland forest (Taber and Fleenor 2003b). Here we feature those species that are more likely to be observed in the lowland vegetation of relict swamps and marshes.

Swamp spiders
Dolomedes tenebrosus, D. triton
(FIGS. 12-1–12-2)

Biology: These two very large species are better known as nursery web spiders because of the distinctive webs with which they enshroud vegetation and in which the young are guarded, and as fishing spiders because of their diet. The spotted species runs across the surface of ponds and lagoons in broad daylight in Palmetto State Park with the aid of hydrophobic substances coating its own surface underneath (Carico 1973). We saw the dark swamp spider under different circumstances, at night on trees along trails.

Like the whirligig beetles of Rutledge Creek, swamp spiders detect potential meals by sensing vibrations moving through the water. Thus, a liquid surface substitutes for the ambush web, which they do not spin. Swamp spiders are also able to submerge themselves for more than half an hour. Their food consists of fishes, amphibians, and insects both in and

Fig. 12-1
Dark swamp spider
(*Dolomedes tene-brosus*) with uniden-tified harvestman
prey

Fig. 12-2
Spotted swamp spi-der (*Dolomedes tri-ton*) on pond surface

out of the water. Destructive forest tent caterpillars (*Malacosoma disstria*) are known to be among the prey of the spotted species (*D. triton;* Zimmer-mann and Spence 1989). We saw these attractive blue defoliators in an aggregation on an ash tree growing on a pond margin where we also saw the spotted spider.

Distribution: Dark swamp spiders (*D. tenebrosus*) occur from the Atlantic Ocean to a western limit in the Ottine wetlands, as established here. Spot-ted swamp spiders occur from coast to coast with the exception of the desert Southwest.

Remarks: The dark swamp spider is the larger of the two (length = 26 mm), and it is more commonly found at considerable distances from water, on tree trunks where its brown coloration provides splendid camouflage. The spotted species is nearly as large (length = 20 mm). It is believed that the small spiderlings might disperse by ballooning from their nurseries in the vegetation, but this remained conjectural when the standard reference was published (Carico 1973).

Similar species: Three additional swamp spider species occur in Texas, but we did not notice any. Their colorations and patterns are distinct from those of the two species treated here.

Golden garden spider
Argiope aurantia
(FIG. 12-3)

Biology: These are very large, orb-weaving spiders that do not hide during the day in a retreat, as the giant orb weaver does, but remain, typically upside down, in the center of the distinctive web. During late spring and early summer the garden spider's huge silk snares are confronted at every turn in bushes and shrubs along Rutledge Creek and the San Marcos River. Here they catch and consume grasshoppers in great quantity, including the pestiferous differential grasshopper (*Melanoplus differentialis*). Females spin large, round, papery egg sacs that readily identify the species responsible even when the adults of both sexes have passed with the seasons.

Fig. 12-3
Golden garden spider (*Argiope aurantia*)

Distribution: The golden garden spider and its relative discussed in the next section occur from coast to coast within the United States.

Remarks: Length = 28 mm. One visitor to Palmetto State Park referred to this species in casual conversation as the banana spider because of its yellow and black color and perhaps its size and shape as well.

Similar species: Adult golden garden spiders have a wide, dark band extending from front to back on the upper surface of the abdomen. The banded garden spider (*A. trifasciata*) lacks this band but has a series of stripes on the abdomen, which extend from side to side, and it has more black bands on the legs. Its egg sac is not rounded but is shaped like a kettle drum (Kaston 1978).

Giant orb weaver
Araneus bicentenarius
(FIG. 12-4)

Biology: Big adult females appear briefly in spring. Their webs are nearly circular and grow to diameters greater than those of the garden spiders, though the latter are larger animals. Some of these snares are several feet wide. During the day the spider hides in a curled leaf (the "refugium") near one of the top edges of its web. A remarkable feature is the lichenlike pattern on the abdomen, which provides protective coloration on lichen-bearing trees.

Distribution: From the Atlantic Ocean to a western extreme near Austin, Texas, not far north of the Ottine swamps.

Remarks: Length = 25 mm. The literature describes this orb weaver as rare

Fig. 12-4
Giant orb weaver
(*Araneus bicentenarius*)

(Levi 1971). In wild parts of central Texas adults are briefly and season-
ally common.

Similar species: None.

Fig. 12-5
Marbled orb weaver
(*Araneus
marmoreus*) from
North Soefje Marsh

Marbled orb weaver
Araneus marmoreus
(FIG. 12-5)

Biology: We saw a single specimen of this attractive species during years of
study in central Texas forests and wetlands. It was on a Sesbania shrub at
the edge of a swamp. These spiders build webs of the same general type as
the garden spider, but they are more likely to use a retreat or "refugium"
as a hiding place and as a safe spot to carry and eat prey caught elsewhere
in the web. The marbled orb weaver falls prey in turn to the organ-pipe
mud-dauber wasp (*Trypoxylon politum*), which stores the spiders as provi-
sion for its grubs (Rehnberg 1987).

Distribution: Widely distributed across the Northern Hemisphere, this
species, unlike most in the wetlands, is not confined to the New World.

Remarks: Length = 18 mm. The black and yellow coloration of our swamp
specimen is unlike any pattern that we saw in the literature (Levi and
Levi 1987; Blanke and Merklinger 1982). One brightly colored variant is
known as the Halloween spider because of its resemblance to a jack-o'-
lantern.

Similar species: None. We thank Herb Levi for identifying this animal.

Fig. 12-6
Southern house spider (*Kukulcania hibernalis*); male.

Fig. 12-7
Southern house spider; female with unidentified prey.

Southern house spider
Kukulcania hibernalis
(FIGS. 12-6–12-7)

Biology: The large, dark females might be confused with distantly related tarantulas, though they are not so hairy nor are their habits and habitats similar. We found them among the eaves, cracks, and crevices of the old water tower in Palmetto State Park. On one occasion we found a female under a peeling flake of boxelder trunk. The ugly, irregular web surrounds a central tunnel through which the spider darts to subdue prey that become tangled in the outer threads (Comstock 1940). Prey includes insects as large as the giant walkingstick and the true katydid.

Distribution: From the Atlantic Ocean to at least as far west as central Texas.

Remarks: Length = 19 mm. It is said that the southern house spider "is rarely seen except by the collector" (Comstock 1940, 295). In Palmetto State Park they may be readily seen by passersby after dark. Even the slightest disturbance sends the animals scurrying into their retreat and out of sight.

Similar species: The light brown male might be confused with the brown recluse (*Loxosceles reclusa*), but the house spider is larger, and the male is said not to spin a web. Recluses are uncommon in the wetlands, though they are abundant in the drier uplands of the nearby Lost Pines forest.

Spitting spider
Scytodes undetermined species
(FIG. 12-8)

Biology: This unique predator lives on the stone tower in Palmetto State Park alongside the southern house spider. Spitting spiders earn their name with an unusual manner of capturing prey. They shoot a venomous substance from their fangs in a zigzag squirt that requires only a fraction of a second to do its work (McAlister 1960). This pins the prey to the wood

Fig. 12-8
Spitting spider (*Scytodes* undetermined species)

surface and allows the spider to make its kill without a web and without biting its victim, although it often does that too.

Distribution: Unknown.

Remarks: Length = 6 mm. Our inability to identify this interesting animal to the species level is due to a lack of published work on spitting spiders in general (Jackman 1997).

Similar species: The habitat and appearance of the harmless spitting spider predispose it to misidentification as a brown recluse or possibly as the male southern house spider, which lives alongside. The photos provided here of the spitting spider and southern house spider will obviate that problem.

Fig. 12-9
Giant wolf spider (*Hogna helluo*) with egg sac

Giant wolf spider
Hogna helluo
(FIG. 12-9)

Biology: Giant wolf spiders are abundant beneath decomposing logs during daylight hours, and after dark they move into the open in pursuit of prey and mates. They are active hunters that do not catch their food in a web, nor do they live in a web, though silk is used to some extent when females make their burrows in soil. These are sometimes capped off with a turret of silk and debris. Females carry the large egg sac at the back end of the body rather than with the jaws, and the young spiderlings upon hatching ride about on the mother's body in large numbers.

Distribution: From the Atlantic Ocean to the Rocky Mountains.
Remarks: Length = 38 mm, excluding the legs, making this the largest wolf
spider we saw in the wetlands.
Similar species: The large size of the adults in combination with the color
pattern shown in Fig. 12-9 serves to distinguish the species from close relatives that might occur in the area.

Arrowhead orbweaver
Micrathena sagittata
(FIG. 12-10)

Biology: This colorful red, yellow, and black spider prefers to spin its web
several feet aboveground near the top of a shrub or bush. It takes a wide
variety of prey, including flies, wasps, beetles, and bugs (Jackman 1997).
Distribution: From the Atlantic Ocean to Texas.
Remarks: Length = 9 mm. Two large spines at the rear and a narrow waist
region combine to suggest the shape of a small arrowhead. Though much
smaller than the garden spider and sharing a similar niche, it is never so
abundant.
Similar species: Closely related, much more abundant, and not easy to con-
fuse is the spined orbweaver, *M. gracilis* (length = 10 mm; Fig. 12-11). A
third spined, common, and colorful orbweaver, *Gasteracantha cancriformis*
(length = 10 mm), is distinguished by its crablike body that is broad
rather than elongate. It may be largely white, yellow, or red.

Fig. 12-10
Arrowhead orb-
weaver (*Micrathena
sagittata*)

Fig. 12-11
Spined orbweaver
(*Micrathena gracilis*)

Fig. 12-12
Bark scorpions (*Centruroides vittatus*)

Bark scorpion
Centruroides vittatus
(FIG. 12-12)

Biology: The bark scorpion is a nocturnally active predator that feeds on insects and spiders, which it captures with its claws and subdues with its sting. In the nearby Lost Pines forest we saw it hunting after dark at eye level on tree trunks and even crawling along branches high aboveground. When not seeking prey, it hides beneath logs and snags and under the loose bark of the same. When exposed to light, it scurries beneath the closest available cover. The sting of this species is painful but not life-threatening except perhaps to those few people who are allergic to the venom.

Distribution: From the Mississippi River to the Rio Grande near El Paso, Texas.

Remarks: Length = 60 mm. This is the only scorpion we saw in the wet-lands, and it is far more common in higher, drier habitats. Another common name is the striped scorpion.

Similar species: None.

13
Snails and Slugs

All of the invertebrates featured in this chapter are molluscs, moisture-loving, legless species that thrive in the vicinity of swamps and marshes. After the arthropods, the molluscs compose the most diverse, or "speciose," phylum of invertebrates, though there are relatively few species in these habitats. Two molluscan groups are represented in and near the Ottine wetlands. These are the gastropods (land and water snails and slugs) and the freshwater clams. No snail or slug species is endemic to the Ottine region, but one subspecies, the Palmetto pill snail (*Euchemotrema leai cheatumi*) was made known to science from specimens collected in Palmetto State Park. Clams are not treated here because they are more closely associated with the San Marcos River itself than with the mostly acidic wetlands of the waterway's floodplain.

Acid waters and soils are not the only challenges faced by molluscs living here. Shell makers in particular must deal with a soil type richer in sand than in the limestone material better suited for building shells. Exceptions are those wetlands closest to the river, where flooding of the banks deposits sediment from upstream and, in particular, from other soil types that are more useful to shell makers than sand. Palmetto State Park is the most conspicuous example of such a location.

Water snails
Stenophysa maugeriae, Physa undetermined species
(FIG. 13-1)

Biology: The biologies of these two species appear to be unknown.
Distribution: The tawny aplexa snail (*S. maugeriae*) is a Mexican species that has been introduced into Texas. At the time of writing we could find no

Fig. 13-1
Tawny aplexa snail
(*Stenophysa mauge-riae*)

reliable key to the species of the genus *Physa,* and thus we cannot report a
species distribution for this snail.

Remarks: Length of both species = 20 mm. One close relative of the
unidentified bladder snail that we collected is *Physella virgata,* which lays
up to two hundred eggs after mating. Though the snails often live in
small, readily evaporated ditches, they require several months to reach
adulthood. This species and its relatives are notable for tolerance of polluted waters.

Similar species: None to our knowledge. Identification of small aquatic
snails such as these can be difficult because it often requires the removal
of the animal from its shell and a microscopic study of the soft anatomy.

Land snails
*Rumina decollata, Euchemotrema leai cheatumi, Mesodon roemeri, Rab-
dotus dealbatus, Helicina orbiculata, Anguispira alternata*
(FIGS. 13-2–13-7)

Biology: All of these snails occur beneath decomposing logs, branches, and
leaf litter, and some feed upon fungi. Only the orb-shelled species (*H.
orbiculata*) possesses the familiar lid with which to close the opening to the
shell in time of need. The others probably prevent desiccation during dry
spells by secreting a mucous film at the entrance to the shell. This
"epiphragm" can be thought of as a temporary lid. Rains bring moisture
that dissolves the epiphragm and revives the snail to activity.

The decollate snail (*Rumina decollata*) is slender and spire-shaped, even

Fig. 13-2
Predatory decollate snail (*Rumina decollata*)

Fig. 13-3
Palmetto pill snail (*Euchemotrema leai cheatumi*) on dwarf palmetto

Fig. 13-4
Roemer's snail (*Mesodon roemeri*) on dwarf palmetto

Fig. 13-5
Spire-shelled snail (*Rabdotus dealbatus*) on dwarf palmetto

Fig. 13-6
Orb-shelled snail (*Helicina orbiculata*) beneath the larger and darker Roemer's snail

Fig. 13-7
Checkered snail (*Anguispira alternata*) crawling on a hackberry trunk

after it breaks the tip off its own shell, leaving the characteristic blunted end. It attacks and kills other snails. We found a single specimen of this exotic species in the Gulf cordgrass marsh of Palmetto State Park. The Palmetto pill snail (*E. leai cheatumi*) is abundant beneath logs and stones in swamps and marshes. Like the others, it moves into the open after rains and floods. This particular subspecies was made known to science upon its discovery in Palmetto State Park in 1971 (Fullington 1974).

The most apparent snail in the Ottine wetlands is Roemer's snail (*M. roemeri*). On warm nights following rains it appears in such numbers in Palmetto State Park that one grows accustomed to the crunching of shells underfoot. This event provides an opportunistic feast for bombardier beetles, pillbugs, and other nocturnal hunters and scavengers. In Florida the species is said to require rather moist hardwood habitats near streams (Clench 1954).

The spire-shelled snail (*Rabdotus dealbatus*) is most notable in spring, when it crawls several feet aboveground on bare dead stems and hackberry trunks, a scenario that one would expect in more tropical climes than south-central Texas. It has been said that the species does not remain aboveground in the southern part of the state during the day except in wet weather (Hubright 1960). Nevertheless, in Palmetto State Park we witnessed many exceptions to this behavior.

Remains of the animal are sometimes found by the thousands in ancient dumps used by American Indians in Texas. Several explanations have been suggested. For example, they might have been used for food or for decoration. The ideas could be tested by examining coprolites, the ancient remains of human solid waste, for the presence of snail radulas. This tough part of the snail, itself ironically used in the feeding process, might be preserved after having passed through the human body without being digested (Clark 1973).

The orb-shelled snail is not as commonly encountered as the others. It may be found in wet areas beneath leaves and logs, but we seldom saw it in the open. Checkered snails (*A. alternata*) may sometimes be seen crawling at night on tree trunks in the company of all the species treated here except the two elongate, spirelike snails.

Distribution: The decollate snail is an Old World native that occurs within the United States from the Atlantic Ocean to Texas, as do the native pill snail, the orb-shelled snail, and the checkered snail. Roemer's snail is nearly a Texas endemic, but it does occur in one or two counties in southern Oklahoma. Spire-shelled snails have an unusual range, from Kentucky to New Mexico.

Remarks: Shell size of *Rumina decollata* = 45 mm; of *E. leai cheatumi* = 9 mm; of *M. roemeri* = 24 mm; of *Rabdotus dealbatus* = 26 mm; of *H. orbiculata* = 8.5 mm; of *A. alternata* = 30 mm. The prevalence of ground beetles, such as the purple ground beetle, the bombardier beetles, and the false bombardier beetle, is tied to the abundance of the land snails that they kill and consume.

Similar species: These are the only land snails we encountered or at least identified. An unpublished manuscript by W. L. Pratt Jr. of the Fort Worth Museum of Science lists the following additional species from Palmetto State Park:

Bottleneck snaggletooth snail (*Gastrocopta contracta*)
Southern pinecone snail (*Strobilops texasianus*)
Forshey's ambersnail (*Succinea forsheyi*)
Southeastern tigersnail (*Anguispira strongyloides*)
Carved glyph snail (*Glyphyalinia paucilirata*)
Brittle button snail (*Mesomphix friabilis*)
Minute gem snail (*Hawaiia minuscula*)
Striate wolfsnail (*Euglandina singleyana*)
Prairie rabdotus snail (*Rabdotus mooreanus mooreanus*)
Texas liptooth snail (*Polygyra texasiana*)
Gulf Coast liptooth snail (*Polygyra auriformis*)
Sandyland shrubsnail (*Praticolella pachyloma*)

Carolina slug
Philomycus carolinianus
(FIG. 13-8)

Biology: The Carolina slug lives beneath the cover of bark and within the tunnels bored in logs by insects. We saw them in summer hidden deep within rotting pecan logs. We observed an exposed individual as it fed upon fungus, which is known to be its usual diet (Runham and Hunter 1970). Unlike the garden snails and slugs, many of which are introduced exotics, this is one species that avoids urban and agricultural settings in favor of forested areas.

Distribution: The Carolina slug is an eastern species that prefers humid forests and probably meets its western limit near the Ottine swamps. In this regard it is notable that a checklist of slugs and snails occurring in Travis County just to the north of the wetlands contains no reference to it (Neck 1994).

Fig. 13-8
Carolina slug
(*Philomycus carolinianus*) feeding on
fungus

Remarks: Length = 100 mm. Though quite a few familiar slugs are introduced species, this one is native to the United States.

Similar species: One or two smaller species known as black slugs probably occur in the wetlands, although we did not see them.

Appendix 1
Texas-Endemic Invertebrates of the Ottine Wetlands

1. Texas ash borer (*Anelaphus niveivestitus*): endemic to south and southeast Texas

2. Texas checkered beetle (*Chariessa texana*): previously known only from northwest Texas

3. Texas tortoise beetle (*Coptocycla texana*): endemic to south and southeast Texas

4. Palmetto camel cricket (*Ceuthophilus* new species): endemic to central Texas and not yet bearing a scientific name

5. Uhler's virtuoso katydid (*Amblycorypha uhleri*): endemic to central Texas

6. Orange millipede (*Eurymerodesmus dactylocyphus*): endemic to central Texas

Future work may discover some of these species in northern Mexico.

Appendix 2
Exotic Invertebrates of the Ottine Wetlands

1. Hyacinth glider dragonfly (*Miathyria marcella*): native to South America
2. Smokybrown cockroach (*Periplaneta fuliginosa*): probably native to the Old World
3. Cuban cockroach (*Panchlora nivea*): native to tropical America
4. Southern mole cricket (*Scapteriscus borellii*): native to southern South America
5. Yellow mealworm beetle (*Tenebrio molitor*): widely distributed grain pest
6. Red imported fire ant (*Solenopsis invicta*): native to South America
7. European pillbug (*Armadillo vulgaris*): native to Europe
8. Waxy woodlouse (*Metoponorthus pruinosus*): native to Europe
9. Tawny aplexa snail (*Stenophysa maugeriae*): native to Mexico
10. Decollate snail (*Rumina decollata*): native to the Old World

Glossary

alluvial originating as a deposit left behind by flowing water. Mud deposited on high ground by the flooding San Marcos River is alluvial in origin.

aposematic colors of a kind that warn potential enemies that an individual is poisonous, capable of defending itself, and so on. Red and black is a common aposematic combination.

artesian well a well that flows because of natural pressure exerted by the weight of overlying water.

bog a nutrient-poor peatland that, according to strict usage, receives all of its water from precipitation. This condition arises when peat piles to such a height that the surface no longer has a connection with groundwater or surface water. In one view, bog peat has the further requirement that it must derive primarily from sphagnum mosses. By these criteria there are no true bogs in the Ottine wetlands.

domiciliary occurring in human habitations.

fen a peatland that receives surface water and groundwater and that is nutrient-rich as opposed to the atmospherically watered, nutrient-poor bog.

floodplain the region on either side of a river that is susceptible to being covered by water during floods.

groundwater water originating, most recently at least, from underground. This is the water of seeps, springs, and wells.

lagoon either a shallow pond communicating with a larger body of water or a shallow artificial pond. Combining the two definitions and specifying the San Marcos River as the larger body of water that communicates via flooding, the modified, ephemeral ponds of Palmetto State Park may be considered lagoons.

Malaise trap a tentlike device that intercepts flying insects and collects them in a container at the top of the tent.

marsh a wetland dominated by herbaceous vegetation that grows during at least part of the year from water-covered or water-saturated ground.

mesic hammock relatively high ground within the wetland where soil rich in decomposing vegetation occurs.

mud boil a point in a wetland where gas and/or liquid rises to the surface and escapes from mud in the form of bubbles.

oxbow lake a body of water left behind in the old river bed when a river changes its course.

peat an accumulation of partly decomposed plant material.

peatland land covered by peat but especially those vast tracts of peat-covered land unique to more northern states and countries.

pH a measure of acidity. The lower the pH, the greater the acidity of the water.

quaking a jellylike quivering of certain water-saturated, thickly matted peatlands. Quaking may be induced by jumping up and down in one spot or merely by walking.

seep a place where water reaches the surface from underground while flowing forth at a slow rate.

slough a muddy inlet or creek.

spring a fast-flowing seep.

streamlet groundwater flowing out of the ground at a rate and volume between that of a seep and a spring. We found these flowing cold even in summer in North Soefje Marsh.

surface water the runoff from rain, and, according to one view, the water of rivers and creeks as well.

swamp a wetland dominated by trees that grow during at least part of the year from water-covered ground.

wetland a general term meaning land that, during at least part of the year, is wet enough to support the growth of water-loving vegetation. Water may cover the ground or merely saturate it. Bogs, marshes, peatlands, and swamps are all varieties of wetland.

wet meadow a synonym of "fen" but certainly more descriptive.

Bibliography

Aguiar, A. P. 1997. Mating behavior of *Pelecinus polyturator* (Hymenoptera: Pelecinidae). *Ent. News* 108:117–21.

Alexander, C. P. 1981. Ptychopteridae. Pp. 325–28 in *Manual of Nearctic Diptera, vol. 1.* Ottawa, Ontario, Canada: Biosystematics Research Institute.

Arnett, R. H. 1960. *The beetles of the United States: A manual for identification.* Washington, D.C.: Catholic University of America Press.

———. 2000. *American insects.* 2nd ed. New York: CRC Press.

Arnett, R. H., N. M. Downie, and H. E. Jaques. 1980. *How to know the beetles.* Dubuque, Iowa: Brown.

Atkinson, T. H., P. G. Koehler, and R. S. Patterson. 1991. *Catalog and atlas of the cockroaches (Dictyoptera) of North America north of Mexico.* Entomol. Soc. Am. Misc. Pub. No. 78.

Balduf, W. V. 1945. Bionomic notes on *Menecles insertus* (Say) (Hemiptera, Pentatomidae). *Bull. Brooklyn Ent. Soc.* 40:61–65.

Ball, G. E. 1959. A taxonomic study of the North American Licinini with notes on the Old World species of the genus *Diplocheila* Brullé (Coleoptera). *Mem. Am. Ent. Soc.* 16:1–258 + 15 plates.

Ball, G. E., and A. P. Nimmo. 1983. Synopsis of the species of subgenus *Progaleritina* Jeannel, including reconstructed phylogeny and geographical history (Coleoptera: Carabidae: *Galerita* Fabricius). *Trans. Am. Ent. Soc.* 109:295–356.

Barber, H. G. 1910. Proceedings of the New York Entomological Society meeting held December 7, 1909. *J. New York Ent. Soc.* 18:128–35.

Barber, H. S. 1951. North American fireflies of the genus *Photuris*. *Smith. Misc. Coll.* 117 (1): 1–58.

Beamer, R. H. 1925. Notes on the oviposition of some Kansas cicadas. *Ann. Ent. Soc. Am.* 18:479–82.

———. 1928. Studies on the biology of Kansas Cicadidae. *Univ. Kans. Sci. Bull.* 18:155–263.

Beckemeyer, R. J., and R. E. Charlton. 2000. Distribution of *Microstylum morosum* and *M. galactodes* (Diptera: Asilidae): Significant range extensions. *Ent. News* 111 (2): 84–96.

Bequaert, J. 1957. The Hippoboscidae or louse-flies (Diptera) of mammals and birds.

Part II. Taxonomy, evolution and revision of American genera and species. *Ent. Amer.* 36:417–611.

Biernbaum, C. K. 1989. Distribution and seasonality of branchiopod and malacostracan crustaceans of the Santee National Wildlife Refuge, South Carolina. *Brimleyana* 15:7–30.

Blair, W. F. 1950. The biotic provinces of Texas. *Texas J. Sci.* 2:93–117.

Blanke, V. R., and F. Merklinger. 1982. Die Variabilität von Zeichnungsmuster und Helligkeit des Abdomens bei *Araneus diadematus* Clerck und *Araneus marmoreus* Clerck (Arachnida: Araneae). *Z. zool. Syst. Evolut.-forsch.* 20:63–75.

Blatchley, W. S. 1910. *An illustrated descriptive catalogue of the Coleoptera or beetles (exclusive of Rhynchophora) known to occur in Indiana.* Indianapolis, Ind.: Nature Publishing.

———. 1920. *Orthoptera of northeastern America.* Indianapolis, Ind.: Nature Publishing.

———. 1926. *Heteroptera, or true bugs of eastern North America.* Indianapolis, Ind.: Nature Publishing.

Blatchley, W. S., and C. W. Leng. 1916. *The Rhynchophora or weevils of north eastern America.* Indianapolis, Ind.: Nature Publishing.

Blume, R. R., and A. Aga. 1976. *Phanaeus difformis* LeConte (Coleoptera: Scarabaeidae): Clarification of published descriptions, notes on biology, and distribution in Texas. *Col. Bull.* 30:199–205.

Bogusch, E. R. 1928. Composition and seasonal aspects of the Gonzales County marsh associes. Master's thesis, University of Texas at Austin.

———. 1930. Trend of succession in a southern bog. *Ill. Acad. Sci., Papers in Biology and Agriculture* 23:285–97.

Bohls, S. W. 1944. *The mosquitoes of Texas.* Austin: Texas State Health Department.

Bowles, D. E. 1998. Life history of *Bittacomorpha clavipes* (Fabricius) (Diptera: Ptychopteridae) in an Ozark spring, U.S.A. *Aquat. Insects* 20:29–34.

Boyle, W. W. 1956. A revision of the Erotylidae of America north of Mexico (Coleoptera). *Bull. Am. Mus. Nat. Hist.* 110:1–172.

Briceño, R. D., and W. G. Eberhard. 1995. The functional morphology of male cerci and associated characters in 13 species of tropical earwigs (Dermaptera: Forficulidae, Labiidae, Carcinophoridae, Pygidicranidae). *Smith. Contrib. Zool.* 0 (555): 1–63.

Bromley, S. W. 1931. New Asilidae, with a revised key to the genus *Stenopogon* Loew: (Diptera). *Ann. Ent. Soc. Am.* 24:427–35.

———. 1933. Cicadas in Texas. *Psyche* 40:130.

———. 1934. The robber flies of Texas. *Ann. Ent. Soc. Am.* 27:74–113.

Brown, W. J. 1946. Notes on some species of *Canthon* and *Dichelonyx* (Coleoptera: Scarabaeidae). *Can. Ent.* 78:104–9.

Brues, C. T. 1928. A note on the genus *Pelecinus. Psyche* 35:205–9.

Brushwein, J. R., J. D. Culin, and K. M. Hoffman. 1995a. Development and reproductive behavior of *Mantispa viridis* Walker (Neuroptera: Mantispidae). *J. Ent. Sci.* 30:99–111.

———. 1995b. Seasonal phenology and overwintering of *Mantispa viridis* Walker (Neuroptera: Mantispidae) in South Carolina. *J. Ent. Sci.* 30:112–19.

Brushwein, J. R., K. M. Hoffman, and J. D. Culin. 1992. Spider (Araneae) taxa associated with *Mantispa viridis* (Neuroptera: Mantispidae). *J. Arach.* 20:153–56.

Bryant, V. M., Jr. 1977. A 16,000 year pollen record of vegetational change in central Texas. *Palynology* 1:143–56.

Buchanan, W. D. 1960. Biology of the oak timberworm *Arrhenodes minutus*. *J. Econ. Ent.* 53:510–13.

Buchler, E. R., T. B. Wright, and E. D. Brown. 1981. Functions of stridulation by the passalid beetle *Odontotaenius disjunctus* (Coleoptera: Passalidae). *Animal Behaviour* 29:483–86.

Bullard, F. M. 1935. Geological background. Pp. 18–19 in *First Scientific Field Meet: Palmetto State Park, Ottine, Texas.* Gonzales, Tex.: Palmetto State Park, Tex. Board.

Burgess, A. F., and C. W. Collins. 1917. The genus *Calosoma*. *USDA Bull.* no. 417: 1–124.

Byers, G. W. 1987. Order Mecoptera. Pp. 246–52 in *Immature insects, vol. 1.* Dubuque, Iowa: Kendall/Hunt.

Canterbury, L. E. 1979. Studies of the genus *Sialis* (Sialidae: Megaloptera) in eastern North America. Ph.D. diss., University of Louisville, Louisville, Kentucky.

Carico, J. E. 1973. The Nearctic species of the genus *Dolomedes* (Araneae: Pisauridae). *Bull. Mus. Comp. Zool.* 144 (7): 435–88.

Carpenter, F. M. 1931. Revision of the Nearctic Mecoptera. *Bull. Mus. Comp. Zool.* 72:205–77 + 8 plates.

Carpenter, S. J., and W. J. LaCasse. 1955. *Mosquitoes of North America (north of Mexico).* Berkeley: University of California Press.

Carroll, S. P. 1991. The adaptive significance of mate guarding in the soapberry bug, *Jadera haematoloma* (Hemiptera: Rhopalidae). *J. Insect Behav.* 4:509–30.

Carroll, S. P., and C. Boyd. 1992. Host race radiation in the soapberry bug: Natural history with the history. *Evolution* 46:1052–69.

Carroll, S. P., and H. Dingle. 1996. The biology of post-invasion events. *Biol. Conserv.* 78:207–14.

Cartwright, O. L. 1959. Scarab beetles of the genus *Bothynus* in the United States (Coleoptera: Scarabaeidae). *Proc. U.S. Nat. Mus.* 108 (3409): 515–41.

Caudell, A. N. 1903. The Phasmidae, or walkingsticks, of the United States. *Proc. U.S. Nat. Mus.* 26:863–85 + 4 plates.

Chandler, H. P. 1956. Megaloptera. Pp. 229–33 in *Aquatic insects of California,* ed. R. L. Usinger. Berkeley: University of California Press.

Chelf, C. 1941. Peat bogs in Gonzales County with notes on other bogs. Bureau of Economic Geology, Circular no. 34, University of Texas at Austin.

Cherepanov, A. I. 1988. *Cerambycidae of northern Asia.* Vol. I, *Prioninae, Disteniinae, Lepturinae, Aseminae.* New Delhi, India: Amerind.

Chopra, N. P. 1973. A revision of the genus *Niesthrea* Spinola (Rhopalidae: Hemiptera). *J. Nat. Hist.* 7:441–59.

Chvála, M., L. Lyneborg, and J. Moucha. 1972. *The horse flies of Europe.* Copenhagen, Denmark: Entomological Society of Copenhagen.

Clark, J. W. 1973. The problem of the land snail genus *Rabdotus* in Texas archeological sites. *Nautilus* 87:24.

Clench, W. J. 1954. *Mesodon thyroidus* (Say) in Florida. *Nautilus* 68:23–24.

Coelho, J. 2002. Spurred on to greater depths. *Natural History* 111 (6): 20–22.

Cohn, T. J. 1965. *The arid-land katydids of the North American genus* Neobarrettia *(Orthoptera: Tettigoniidae): Their systematics and a reconstruction of their history.* Misc. Pub. no. 126. Ann Arbor: Museum of Zoology, University of Michigan.

Cole, F. R. 1969. *The flies of western North America.* Berkeley: University of California Press.

Collinge, W. E. 1946. Some observations on the woodlouse *Porcellionides pruinosus* (Brandt). *Ann. & Mag. Nat. Hist.,* ser. 11, vol. 13:359–60.

Collins, M. M., and R. D. Weast. 1961. *Wild silk moths of the United States.* Cedar Rapids, Iowa: Collins Radio.

Comstock, J. H. 1940. *The spider book.* New York: Doubleday, Doran.

Contreras-Ramos, A. 1998. *Systematics of the dobsonfly genus* Corydalus *(Megaloptera: Corydalidae).* Lanham, Md.: Thomas Say Publications in Entomology, Entomological Society of America.

Cooper, W. E., Jr. 1981. Mimicry and spatial occupation in the mydas fly, *Mydas clavatus. J. Ala. Acad. Sci.* 52:58–65.

Covell, C. V., Jr. 1984. *A field guide to the moths of eastern North America.* Boston: Houghton Mifflin.

Craighead, F. C. 1923. *North American Cerambycid larvae.* Technical Bulletin no. 27, n.s., Department of Agriculture, Dominion of Canada.

Cuda, J. P., C. J. DeLoach, and T. O. Robbins. 1990. Population dynamics of *Melipotis indomita* (Lepidoptera: Noctuidae), an indigenous natural enemy of mesquite, *Prosopis* spp. *Envir. Ent.* 19:415–22.

Cumley, R. W. 1931. A geologic section across Caldwell County, Texas. Master's thesis, University of Texas at Austin.

Davis, W. T. 1910. Observations on *Cicada pruinosa* and a description of a new species. *Ent. News* 21:457–58.

———. 1922. An annotated list of the cicadas of Virginia with description of a new species. *J. New York Ent. Soc.* 30:36–52 + 1 plate.

———. 1935. New cicadas with notes on North American and West Indian species. *J. New York Ent. Soc.* 43:173–200.

———. 1944. The remarkable distribution of an American cicada: A new genus, and other cicada notes. *J. New York Ent. Soc.* 52:213–22.

DeLoach, C. J. 1994. Feeding behavior of *Melipotis indomita* (Lepidoptera: Noctuidae), a herbivore of mesquite (*Prosopis* spp.). *Envir. Ent.* 23:161–66.

Dirsh, V. M. 1974. *Genus* Schistocerca *(Acridomorpha, Insecta).* The Hague, The Netherlands: Dr. W. Junk Publishers.

Dolin, P. S., and D. C. Tarter. 1981. Life history and ecology of *Chauliodes rastricornis* Rambur and *C. pectinicornis* (Linnaeus) (Megaloptera: Corydalidae) in Greenbottom swamp, Cabell County, West Virginia. *Brimleyana* 0 (7): 111–20.

Dozier, H. L. 1920. An ecological study of hammock and piney woods insects in Florida. *Ann. Ent. Soc. Am.* 13:325–80.

Duffy, E. A. J. 1960. *A monograph of the immature stages of Neotropical timber beetles (Cerambycidae).* London: Trustees of the British Museum.

Dunkle, S. 1989. *Dragonflies of the Florida peninsula, Bermuda, and the Bahamas.* Gainesville, Fla.: Scientific Publishers.

———. 2000. *Dragonflies through binoculars.* New York: Oxford University Press.

Eads, R. B. 1950. *The fleas of Texas.* Austin: Texas State Health Department.

Eberhard, W. G. 1998. Sexual behavior of *Acanthocephala declivis guatemalana* (Hemiptera: Coreidae) and the allometric scaling of their modified hind legs. *Ann. Ent. Soc. Am.* 91:863–71.

Edwards, J. G. 1949. *Coleoptera or beetles east of the Great Plains.* Ann Arbor, Mich.: Edwards Brothers.

Ellis, R. J. 1961. A life history study of *Asellus intermedius* Forbes. *Trans. Amer. Mic. Soc.* 80:80–102.

Endrödi, 1985. *The Dynastinae of the world.* Boston: Dr. W. Junk Publishers.

Erwin, T. L. 1970. A reclassification of bombardier beetles and a taxonomic revision of the North and Middle American species (Carabidae: Brachinida). *Quaestiones Entomologicae* 6:4–215.

Esselbaugh, C. O. 1948. Notes on the bionomics of some midwestern Pentatomidae. *Ent. Amer.* 28:1–73.

Evans, A. V., and C. L. Bellamy. 1996. *An inordinate fondness for beetles.* New York: Holt.

Evenhuis, N. L., and D. J. Greathead. 1999. *World catalog of bee flies (Diptera: Bombyliidae).* Leiden, Netherlands: Backhuys.

Fattig, P. W. 1933. Food of the robber fly, *Mallophora orcina* (Wied) (Diptera). *Can. Ent.* 65:119–20.

Fitzgerald, T. D. 1995. *The tent caterpillars.* Ithaca, N.Y.: Cornell University Press.

Folkerts, G. W. 1967. Mutualistic cleaning behavior in an aquatic beetle (Coleoptera). *Col. Bull.* 21:27–28.

Forschler, B. T., and G. L. Nordin. 1989. Impact of *Beauveria bassiana* on the cottonwood borer *Plectrodera scalator* Coleoptera Cerambycidae in a commercial cottonwood nursery. *J. Ent. Sci.* 24:186–90.

Forsyth, T. G. 1991. Feeding and locomotory functions in relation to body form in five species of ground beetle (Coleoptera: Carabidae). *J. Zoology (London)* 223:233–63.

Freeman, B. 2003. A fallout of black witches (*Ascalapha odorata*) associated with Hurricane Claudette. *News of the Lepidopterists' Soc.* 45 (3): 71.

Froeschner, R. C. 1952. A synopsis of the Cicadidae of Missouri (Homoptera). *J. New York Ent. Soc.* 60:1–14.

Fullington, R. W. 1974. Two new land gastropods from Texas (*Zonitoides* and *Stenotrema*). *Nautilus* 88 (4): 91–93.

Gangwere, S. K. 1990. Food selection and feeding behavior in the species of *Neobarrettia* Rehn, 1901, a New World genus of predacious katydid (Orthoptera: Tettigoniidae). *Bol. San. Veg. Plagas (Fuera de serie)* 20:291–98.

Garthwaite, R., and C. Sassaman. 1985. *Porcellionides floria*, new species, from North America: Provinciality in the cosmopolitan isopod *Porcellionides pruinosus* (Brandt). *J. Crustacean Biology* 5:539–55.

Gibbons, J. R. H. 1979. A model for sympatric speciation in *Megarhyssa* (Hymenoptera: Ichneumonidae): Competitive speciation. *Amer. Nat.* 114:719–41.

Gibson, E. H., and A. Holdridge. 1918. Notes on the North and Central American species of *Acanthocephala* Lap. (Fam. Coreidae: Heteroptera). *Can. Ent.* 50:237–41.

Gibson, W. W. 1965. An observation on the oviposition habits of *Mydas clavatus* (Diptera: Mydaidae). *J. Kans. Ent. Soc.* 38:196–97.

Gidaspow, T. 1959. North American caterpillar hunters of the genera *Calosoma* and *Callisthenes* (Coleoptera, Carabidae). *Bull. Am. Mus. Nat. Hist.* 116:229–343.

Goldsmith, S. K., Z. Stewart, S. Adams, and A. Trimble. 1996. Body size, male aggression, and male mating success in the cottonwood borer, *Plectrodera scalator* (Coleoptera: Cerambycidae). *J. Insect Behav.* 9:719–27.

Goodwin, J. T., and B. M. Drees. 1996. The horse and deer flies (Diptera: Tabanidae) of Texas. *Southwest. Entomol. Suppl.* 20:1–140.

Graham, A. 1958. Pollen studies of some Texas peat deposits. Master's thesis, University of Texas at Austin.

Graham, A., and C. Heimsch. 1960. Pollen studies of some Texas peat deposits. *Ecology* 41:751–63.

Gray, I. E. 1946. Observations on the life history of the horned passalus. *Am. Midland Nat.* 35:728–46.

Green, J. W. 1956. Revision of the Nearctic species of *Photinus* (Lampyridae: Coleoptera). *Proc. Cal. Acad. Sci.* 28:561–613.

Halffter, G., V. Halffter, and I. Lopez G. 1974. *Phanaeus* behavior: Food transportation and bisexual cooperation. *Envir. Ent.* 3:341–45.

Halstead, J. A. 1989. On the biology and rarity of the owlfly *Ululodes arizonensis* in California (Neuroptera: Ascalaphidae). *Pan-Pac. Ent.* 65:418–19.

Hamilton, J. 1892. Notes on Coleoptera. No. 9. *Can. Ent.* 24:37–42.

Hancock, J. L. 1916. Pink katydids and the inheritance of pink coloration (Orth.). *Ent. News* 27:70–82.

Harrington, W. H. 1882. Long-stings. *Can. Ent.* 14:81–84.

Hart, C. A. 1895. On the entomology of the Illinois River and adjacent waters. *Bull. Illinois State Laboratory of Natural History* 4:149–284.

Hartigan, P., and G. Lasley, comps. 1987. *Birds of Palmetto State Park: A field checklist.* Austin: Resource Management Section, Texas Parks and Wildlife Department.

Harvey, I. F., and S. F. Hubbard. 1987. Observations on the reproductive behaviour of *Orthemis ferruginea* (Fabricius) (Anisoptera: Libellulidae). *Odonatologica* 16:1–8.

Hatch, M. H. 1925. An outline of the ecology of Gyrinidae. *Bull. Brooklyn Ent. Soc.* 20:101–14.

Hebard, M. 1917. Notes on the earwigs (Dermaptera) of North America, north of the Mexican boundary. *Ent. News* 28:311–23.

———. 1934. The Dermaptera and Orthoptera of Illinois. *Bull. Ill. Nat. Hist. Surv.* 20:124–279.

———. 1941. The group Pterophyllae as found in the United States (Tettigoniidae: Pseudophyllinae). *Trans. Am. Ent. Soc.* 67:197–219 + 2 plates.

———. 1943. The Dermaptera and Orthopterous families Blattidae, Mantidae, and Phasmidae of Texas. *Trans. Am. Ent. Soc.* 68:239–319.

Helfer, J. R. 1953. *How to know the grasshoppers, cockroaches, and their allies.* Dubuque, Iowa: Brown.

Henderson, A., G. Galeano, and R. Bernal. 1995. *Field guide to the palms of the Americas.* Princeton, N.J.: Princeton University Press.

Henderson, L. S. 1939. A revision of the genus *Listronotus:* I (Curculionidae: Coleoptera). *Univ. Kans. Sci. Bull.* 26:215–337.

Hickin, N. E. 1958. *Romaleum rufulum* Hald. (Col. Cerambycidae) in American oak. *Ent. Monthly Mag.,* October, 233.

Hildebrand, E. F. 1935. History of Palmetto State Park. Pp. 2–4 in *First Scientific Field Meet: Palmetto State Park, Ottine, Texas.* Gonzales, Tex.: Palmetto State Park, Tex. Board.

Hinds, W. E. 1901. Strength of *Passalus cornutus* Fab. *Ent. News* 12:257–62 + 1 plate.

Hobbs, H. H., Jr. 1989. An illustrated checklist of the American crayfishes (Decapoda: Astacidae, Cambaridae, and Parastacidae). *Smith. Contrib. Zool.* 480:1–236.

Hobbs, H. H., Jr., and H. H. Hobbs III. 1990. A new crayfish (Decapoda: Cambaridae) from southeastern Texas. *Proc. Biol. Soc. Wash.* 103:608–13.

Hopkin, S. P., and H. J. Read. 1992. *The biology of millipedes.* New York: Oxford University Press.

Howard, D. F., M. S. Blum, T. H. Jones, and D. W. Phillips. 1982. Defensive adaptations of eggs and adults of *Gastrophysa cyanea* (Coleoptera: Chrysomelidae). *J. Chem. Ecol.* 8:453–62.

Howard, L. O. 1904. *The insect book.* New York: Doubleday, Page.

Hubbard, C. A. 1947. *Fleas of western North America.* Ames: Iowa State College Press.

Hubbell, T. H. 1936. *A monographic revision of the genus* Ceuthophilus. Biological Science Series, vol. II, no. 1. Gainesville: University of Florida Publications.

———. 1960. The sibling species of the *alutacea* group of the bird-locust genus *Schistocerca* (Orthoptera, Acrididae, Cyrtacanthacridinae). *Misc. Pub. Mus. Zoology,* no. 116. Ann Arbor: University of Michigan.

Hubright, L. 1960. The genus *Bulimulus* in southern Texas. *Nautilus* 74:68–70.

Hudson, W. G. 1987. Ontogeny of prey selection in *Sirthenea carinata:* Generalist juveniles become specialist adults. *Entomophaga* 32:399–406.

Hull, F. M. 1973. *Bee flies of the world.* Washington, D.C.: Smithsonian Institution Press.

Huner, J. V. 2002. *Procambarus* spp. Pp. 541–84 in *Biology of freshwater crayfish,* ed. D. M. Holdich. Oxford, England: Blackwell Science.

Hurd, P. D., Jr. 1959. Beefly [*sic*] parasitism of the American carpenter bees belonging to the genus *Xylocopa* Latreille. *J. Kans. Ent. Soc.* 32 (2): 53–58.

Ideker, J. 1979. Adult *Cybister fimbriolatus* are predaceous (Coleoptera: Dytiscidae). *Col. Bull.* 33:41–44.

Isely, F. B. 1941. Researches concerning Texas Tettigoniidae. *Ecol. Mon.* 11:457–75.

————. 1944. Correlation between mandibular morphology and food specificity in grasshoppers. *Ann. Ent. Soc. Am.* 37:47–67.

Jackman, J. A. 1997. *A field guide to spiders & scorpions of Texas.* Houston, Tex.: Gulf Publishing.

Jacques, R. L., Jr. 1988. *The potato beetles. Flora and fauna handbook no. 3.* New York: Brill.

Jensen, M. N. 2000. Silk moth deaths show perils of biocontrol. *Science* 290:2230–31.

Johnson, C. 1988. Species identification in the eastern *Crematogaster* (Hymenoptera: Formicidae). *J. Ent. Sci.* 23:314–32.

Johnson, G. H. 1972. Flight behavior of the predaceous diving beetle, *Cybister fimbriolatus fimbriolatus* (Say) (Coleoptera: Dytiscidae). *Col. Bull.* 26:23–24.

Johnson, G. H., and W. Jakinovich, Jr. 1970. Feeding behavior of the predaceous diving beetle *Cybister fimbriolatus fimbriolatus* (Say). *Bioscience* 20:1111.

Johnson, N. F., and L. Musetti. 1999. Revision of the proctotrupoid genus *Pelecinus* Latreille (Hymenoptera: Pelecinidae). *J. Nat. Hist.* 33:1513–43.

Jones, W. A., Jr., H. E. Walker, P. C. Quimby, and J. D. Ouzts. 1985. Biology of *Niesthrea louisianica* (Hemiptera: Rhopalidae) on selected plants, and its potential for biocontrol of velvetleaf *Abutilon theophrasti* (Malvaceae). *Ann. Ent. Soc. Am.* 78:326–30.

Kaston, B. J. 1978. *How to know the spiders.* 3rd ed. Dubuque, Iowa: Brown.

King, E. A., Jr. 1961. Geology of northwestern Gonzales County. Master's thesis, University of Texas at Austin.

King, J. L. 1919. Notes on the biology of the carabid genera *Brachynus, Galerita,* and *Chlaenius. Ann. Ent. Soc. Am.* 12:382–90.

Kirn, A. J. 1935. Birds of the Ottine area. Pp. 12–14 in *First Scientific Field Meet: Palmetto State Park, Ottine, Texas.* ?Gonzales, Tex.: Palmetto State Park, Tex. Board.

Knetzger, A. 1908. Doings of societies. *Ent. News* 19:497–98.

Knudson, E. C. 1986. New species of olethreutine moths (Tortricidae) from Texas and Louisiana. *J. Lepidopterists' Society* 40:322–26.

Krombein, K. V. 1938. Studies in the Tiphiidae (Hymenoptera: Aculeata). II. A revision of the Nearctic Myzininae. *Trans. Am. Ent. Soc.* 64:227–92.

Lago, P. K., and M. O. Mann. 1987. Survey of Coleoptera associated with flowers of wild carrot *Daucus carota* L. Apiaceae in northern Mississippi. *Col. Bull.* 41:1–8.

Larson, D. A., V. M. Bryant, and T. S. Patty. 1972. Pollen analysis of a central Texas bog. *Amer. Midl. Nat.* 88:358–67.

LeConte, J. L. 1881. Synopsis of the Lampyridae of the United States. *Trans. Ent. Soc. Am.* 9:15–72.

Leng, C. W. 1902. Revision of the Cicindelidae of boreal America. *Trans. Am. Ent. Soc.* 28:93–186 + 3 plates.

Lent, H., and P. Wygodzinsky. 1979. Revision of the Triatominae (Hemiptera: Reduviidae) and their significance as vectors of Chagas' disease. *Bull. Am. Mus. Nat. Hist.* 163:123–520.

Levi, H. W. 1971. The *diadematus* group of the orb-weaver genus *Araneus* north of Mexico (Araneae: Araneidae). *Bull. Mus. Comp. Zool.* 141:131–79.

Levi, H. W., and L. R. Levi. 1987. *Spiders and their kin.* New York: Golden Press.

Lewis, W. M., Jr. 2001. *Wetlands explained.* Oxford: Oxford University Press.

Linsley, E. G. 1961. *The Cerambycidae of North America. Part I.* Berkeley: Univ. Calif. Pub. Entomol. 18.

———. 1962a. *The Cerambycidae of North America. Part II.* Berkeley: Univ. Calif. Pub. Entomol. 19.

———. 1962b. *The Cerambycidae of North America. Part III.* Berkeley: Univ. Calif. Pub. Entomol. 20.

———. 1963. *The Cerambycidae of North America. Part IV.* Berkeley: Univ. Calif. Pub. Entomol. 21.

———. 1964. *The Cerambycidae of North America. Part V.* Berkeley: Univ. Calif. Pub. Entomol. 22.

Linsley, E. G., and J. A. Chemsak. 1984. *The Cerambycidae of North America, Part VII, no. 1.* Berkeley: Univ. Calif. Pub. Entomol. 102.

———. 1995. *The Cerambycidae of North America. Part VII, no. 2.* Berkeley: Univ. Calif. Pub. Entomol. 114.

———. 1997. *The Cerambycidae of North America, Part VIII.* Berkeley: Univ. Cal. Pub. Entomol. 117.

Linsley, E. G., and P. D. Hurd, Jr. 1959. The larval habits of *Plinthocoelium suaveolens plicatum* (LeConte) (Coleoptera: Cerambycidae). *Bull. Southern Cal. Acad. Sciences* 58:27–33.

Linsley, E. G., J. N. Knull, and M. Statham. 1961. *A list of Cerambycidae from the Chiricahua Mountain area, Cochise County, Arizona (Coleoptera).* New York: American Museum Novitates no. 2050.

Linsley, E. G., and J. O. Martin. 1933. Notes on some longicorns from subtropical Texas (Coleop.: Cerambycidae). *Ent. News* 44:178–83.

Lockett, L. 2003. The native presence of *Sabal mexicana (Sabal texana)* north of the lower Rio Grande Valley. *NPSOT News* 21 (3): 1, 6–10.

Lodwick, L. N., and J. A. Snider. 1980. The distribution of *Sphagnum* taxa in Texas. *The Bryologist* 83:214–18.

Maa, T. C. 1969. Further notes on Lipopteninae (Diptera: Hippoboscidae). *Pacific Insects Monograph* 20:205–36.

Maa, T. C., and B. V. Peterson. 1987. Hippoboscidae. Pp. 1271–81 in *A manual of Nearctic Diptera, vol. 2,* edited by J. F. McAlpine, B. V. Peterson, G. E. Shewell, H. J. Teskey, J. R. Vockeroth, and D. M. Wood. Ottawa, Ontario: Research Branch, Agriculture Canada.

Macgown, J., and M. Macgown. 1996. Observation of a nuptial flight of the horned passalus beetle, *Odontotaenius disjunctus* (Illiger) (Coleoptera: Passalidae). *Col. Bull.* 50:201–3.

MacRoberts, B. R., and M. H. MacRoberts. 1998. Floristics of muck bogs in east central Texas. *Phytologia* 85:61–73.

Manee, A. H. 1908. Some observations at Southern Pines, N. Carolina. *Ent. News* 19:286–89.

Marston, N. 1970. Revision of New World species of *Anthrax* (Diptera: Bombyliidae), other than the *Anthrax albofasciatus* group. *Smith. Contrib. Zool.* 43:1–148.

Martin, J. W., and D. Belk. 1988. Review of the clam shrimp family Lynceidae Stebbing, 1902 (Branchiopoda: Conchostraca), in the Americas. *J. Crust. Biol.* 8:451–82.

Martin, J. W., B. E. Felgenhauer, and L. G. Abele. 1986. Redescription of the clam shrimp *Lynceus gracilicornis* (Packard) (Branchiopoda, Conchostraca, Lynceidae) from Florida, with notes on its biology. *Zoologica Scripta* 15:221–32.

Maxwell, R. A. 1970. Palmetto State Park. Pp. 150–53 in *Geologic and historic guide to state parks of Texas.* Austin: University of Texas, Bureau of Economic Geology—Guidebook no. 10.

McAlister, W. H. 1960. The spitting habit in the spider *Scytodes intricata* Banks (Family Scytodidae). *Texas J. Sci.* 12:17–20.

McDermott, F. A. 1910. Note on the light-emission of some American Lampyridae. *Can. Ent.* 42:357–64.

———. 1967. The North American fireflies of the genus *Photuris* DeJean: A modification of Barber's key (Coleoptera: Lampyridae). *Col. Bull.* 21:106–16.

McFadden, M. W. 1967. Soldier fly larvae in America north of Mexico. *Proc. U.S. Nat. Mus.* 121 (3569): 1–72.

McPherson, J. E. 1982. *The Pentatomoidea (Hemiptera) of northeastern North America.* Carbondale: Southern Illinois University Press.

Miller, R. H., and G. N. Cameron. 1987. Effects of temperature and rainfall on populations of *Armadillidium vulgare* (Crustacea: Isopoda) in Texas. *Am. Midland Nat.* 117:192–98.

Mitchell, J. D., and W. D. Pierce. 1911. The weevils of Victoria County, Texas. *Proc. Ent. Soc. Wash.* 13:45–62.

Mitsch, W. J., and J. G. Gosselink. 2000. *Wetlands.* 3rd ed. New York: Wiley.

Mohlenbrock, R. H. 2002. Going with the flow: A Texas river winds through town and country. *Natural History* 111 (1): 14–15.

Moldenke, A. R. 1970. *A revision of the Clytrinae of North America north of the Isthmus of Panama (Coleoptera: Chrysomelidae).* Stanford, Calif.: Stanford University Press.

Muchmore, W. B. 1990. Terrestrial Isopoda. Pp. 805–17 in *Soil biology guide,* edited by D. L. Dindal. New York: Wiley.

Nagel, M. G., and W. H. Cade. 1983. On the role of pheromones in aggregation formation in camel crickets, *Ceuthophilus secretus* (Orthoptera: Gryllacrididae). *Can. J. Zool.* 61:95–98.

Neck, R. W. 1994. Land snails of Travis County. Pp. 146–51 in *Birds and other wildlife of south central Texas,* edited by E. A. Kutac and S. C. Caran. Austin: University of Texas Press.

Needham, J. G., and M. J. Westfall, Jr. 1954. *A manual of the dragonflies of North America (Anisoptera).* Berkeley: University of California Press.

Needham, J. G., and M. L. May. 2000. *Dragonflies of North America.* Rev. ed. Gainesville, Fla.: Scientific Publishers.

Nelson, J. W. 1986. Ecological notes on male *Mydas xanthopterus* (Loew) (Diptera: Mydidae) and their interactions with *Hemipepsis ustulata* Dahlbohm (Hymenoptera: Pompilidae). *Pan-Pac. Ent.* 62:316–22.

Nixon, E. S., L. F. Chambless, and J. L. Malloy. 1973. Woody vegetation of a palmetto [*Sabal minor* (Jacq.) Pers.] area in east Texas. *Texas J. Sci.* 24:535–41.

O'Brien, C. W. 1981. The larger (4.5+ mm) *Listronotus* of America, north of Mexico (Cylindrorhininae, Curculionidae, Coleoptera). *Trans. Am. Ent. Soc.* 107:69–123.

O'Brien, C. W., and G. J. Wibmer. 1982. Annotated checklist of the weevils (Curculionidae sensu lato) of North America, Central America, and the West Indies (Coleoptera: Curculionoidea). *Mem. Am. Ent. Inst.* 34:1–382.

Opinion. 1897. *Bull. Zoological Nomenclature* 55 (1998): 124–28.

Opler, P. A., and G. O. Krizek. 1984. *Butterflies east of the Great Plains.* Baltimore, Md.: Johns Hopkins University Press.

Packard, A. S. 1882. Larvae of a fly in a hot spring in Colorado. *Amer. Nat.* 16:599–600.

Paris, O. H. 1963. The ecology of *Armadillidium vulgare* (Isopoda: Oniscoidea) in California grassland: Food, enemies, and weather. *Ecol. Mon.* 33:1–22.

Parks, H. B. 1935a. Plant life of Ottine. Pp. 5–11 in *First Scientific Field Meet: Palmetto State Park, Ottine, Texas.* ?Gonzales, Tex.: Palmetto State Park, Tex. Board.

———. 1935b. Amphibia and reptiles. Pp. 14–16 in *First Scientific Field Meet: Palmetto State Park, Ottine, Texas.* ?Gonzales, Tex.: Palmetto State Park, Tex. Board.

———. 1935c. Butterflies of Ottine area. Pp. 16–18 in *First Scientific Field Meet: Palmetto State Park, Ottine, Texas.* ?Gonzales, Tex: Palmetto State Park, Tex. Board.

Patty, T. S. 1968. Pollen analysis and chronology of a central Texas peat bog. Master's thesis, University of Texas at Austin.

Penn, G. H., and H. H. Hobbs, Jr. 1958. A contribution toward a knowledge of the crawfishes of Texas (Decapoda: Astacidae). *Texas J. Sci.* 10:452–83.

Pinto, J. D., and R. B. Selander. 1970. *The bionomics of blister beetles of the genus* Meloe *and a classification of the New World species.* Illinois Biological Monographs no. 42, Urbana, Illinois.

Plummer, F. B. 1941. *Peat deposits in Texas.* Austin: University of Texas Bureau of Economic Geology, Mineral Resource Circular no. 13.

———. 1945. *Progress report on peat deposits in Texas.* Austin: University of Texas Bureau of Economic Geology, Mineral Resource Circular no. 36.

Poinar, G. O., and J. J. Petersen. 1978. *Drilomermis leioderma* n. gen., n. sp. (Mermithidae: Nematoda) parasitizing *Cybister fimbriolatus* (Say). *J. Nemat.* 10:20–23.

Preston-Mafham, K. 1990. *Grasshoppers and mantids of the world.* New York: Facts on File.

Priddle, T. R. 1967. Structures employed by *Actias luna* (Saturniidae) in effecting emergence from the cocoon. *J. Lepidopterists' Society* 21:249–52.

Rasmussen, J. L. 1994. The influence of horn and body size on the reproductive behavior of the horned rainbow scarab beetle *Phanaeus difformis* (Coleoptera: Scarabaeidae). *J. Insect Behav.* 7:67–82.

Raun, G. G. 1958. Vertebrates of a moist, relict area in Texas. Master's thesis, University of Texas at Austin.

———. 1959. Terrestrial and aquatic vertebrates of a moist, relict area in central Texas. *Texas J. Sci.* 11:158–71.

Readio, P. A. 1927. Studies on the biology of the Reduviidae of America north of Mexico. *Univ. Kans. Sci. Bull.* 27:5–291.

Rehn, J. A. G. 1907. A new species of *Ceuthophilus* (Orthoptera) from Kansas. *Ent. News* 18:445–46.

Rehn, J. A. G., and M. Hebard. 1914. Studies in American Tettigoniidae (Orthoptera). I and II. *Trans. Am. Ent. Soc.* 40:271–344 + 4 plates.

Rehnberg, B. G. 1987. Selection of spider prey by *Trypoxylon politum* (Say) (Hymenoptera: Sphecidae). *Can. Ent.* 119:189–94.

Reyes-Castillo, P., and M. Jarman. 1980. Some notes on larval stridulation in Neotropical Passalidae (Coleoptera: Lamellicornia). *Col. Bull.* 34:263–70.

Rice, M. E., and W. B. Peck. 1991. *Mantispa sayi* (Neuroptera: Mantispidae) parasitism on spiders (Araneae) in Texas, with observations on oviposition and larval survivorship. *Ann. Ent. Soc. Am.* 84:52–57.

Richardson, H. 1905. A monograph on the isopods of North America. *Bull. U.S. Nat. Mus.* 54:1–727.

Robinson, M. 1948. A review of the genus *Phanaeus* inhabiting the United States. *Trans. Am. Ent. Soc.* 73:299–305.

Rogers, C. D. 1999. *Birds of Palmetto State Park: A field checklist.* Austin: Natural Resources Program, Texas Parks and Wildlife Department.

Rogers, J. S. 1928. Notes on the biology of *Gnophomyia luctuosa* Osten Sacken, with descriptions of the immature stages. *Ann. Ent. Soc. Am.* 21:398–406.

Ross, H. H. 1937. Studies of Nearctic aquatic insects. I. Nearctic alderflies of the genus *Sialis* (Megaloptera, Sialidae). *Bull. Ill. Nat. Hist. Surv.* 21:56–78.

Rossini, C., A. B. Attygalle, A. González, S. R. Smedley, M. Eisner, J. Meinwald, and T. Eisner. 1997. Defensive production of formic acid (80%) by a carabid beetle (*Galerita lecontei*). *Proc. Nat. Acad. Sci, USA* 94:6792–97.

Rowell, C. M., Jr. 1949. A preliminary report on the floral composition of a *Sphagnum* bog in Robertson County. *Texas J. Sci.* 1 (4): 5–53.

Ruggles, A. G. 1915. Life history of *Oberea tripunctata*. *Swed. J. Econ. Ent.* 8:79–85.

Runham, N. W., and P. J. Hunter. 1970. *Terrestrial slugs.* London: Hutchinson.

Sailer, R. I. 1961. The identity of *Lygaeus sidae* Fabricius, type species of the genus *Niesthrea*. *Proc. Ent. Soc. Wash.* 63:293–99.

Sanborne, M. 1983. Some observations on the behaviour of *Arrhenodes minutus* (Drury) (Coleoptera: Brentidae). *Col. Bull.* 37:106–13.

Sargent, T. D. 1976. *Legion of night: The underwing moths.* Amherst: University of Massachusetts Press.

Say, Thomas. 1825. Descriptions of insects of the families Carabici and Hydrocanthari of Latreille, inhabiting North America. *Trans. Am. Phil. Soc.* 2:1–109.

Schaeffer, C. 1933. Notes on some Hispini and Cassidini and descriptions of new species (Coleoptera, Chrysomelidae). *Pan-Pac. Ent.* 9:103–9.

Schmidt, J. O., and M. S. Blum. 1977. Adaptations and responses of *Dasymutilla occidentalis* (Hymenoptera: Mutillidae) to predators. *Ent. Exp. Appl.* 21:99–111.

Schwardt, H. H. 1932. The life histories of two horse-flies. *Ann. Ent. Soc. Am.* 25:631–37.

Setty, L. R. 1931. The biology of *Bittacus stigmaterus* Say (Mecoptera, Bittacusidae). *Ann. Ent. Soc. Am.* 24:467–84.

———. 1940. Biology and morphology of some North American Bittacidae (Order Mecoptera). *Am. Midland Nat.* 23:257–353.

Shearer, G. K. 1956. *Palmetto State Park, Ottine, Texas.* Austin: Texas State Parks Board.

Shelley, R. M. 1989. Revision of the milliped family Eurymerodesmidae (Polydesmida: Chelodesmidea). *Mem. Am. Ent. Soc.* 37:1–112.

Sherman, F., Jr. 1908. The Panorpidae (scorpion-flies) of North Carolina, with notes on the species. *Ent. News* 19:50–54.

Sites, R. W., and J. T. Polhemus. 1994. Nepidae (Hemiptera) of the United States and Canada. *Ann. Ent. Soc. Am.* 87:27–42.

Siverly, R. E. 1972. *Mosquitoes of Indiana.* Indianapolis: Indiana State Board of Health.

Slater, J. A., and R. M. Baranowski. 1978. *How to know the true bugs.* Dubuque, Iowa: Brown.

Slobodchikoff, C. N. 1973. Behavioral studies of three morphotypes of *Therion circumflexum* (Hymenoptera: Ichneumonidae). *Pan-Pac. Ent.* 49:197–206.

———. 1974. Notes on the biology of *Therion circumflexum* (L.), with a description of the immature stages. *Pan-Pac. Ent.* 50:111–17.

———. 1977. Patterns of variation in wasps of the genus *Therion* (Hymenoptera: Ichneumonidae). *Univ. Cal. Pub. Entomol.* 82.

Smith, R. L., and E. Larsen. 1993. Egg attendance and brooding by males of the giant water bug *Lethocerus medius* (Guerin) in the field (Heteroptera: Belostomatidae). *J. Insect Behav.* 6:93–106.

Somes, M. P. 1916. The Phasmidae of Minnesota, Iowa, and Missouri (Orth.). *Ent. News* 27:269–71.

Stillwell, M. A. 1967. The pigeon tremex, *Tremex columba* (Hymenoptera: Siricidae), in New Brunswick. *Can. Ent.* 99:685–89.

Stone, A. 1930. The bionomics of some Tabanidae (Diptera). *Ann. Ent. Soc. Am.* 23:261–304.

Szafranski, P. 2002. New host plant and distributional records for some *Eburia* Lepeletier & Audinet-Serville (Coleoptera: Cerambycidae) in North America including Mexico. *Pan-Pac. Ent.* 78:66–67.

Taber, S. W. 2000. *Fire ants.* College Station: Texas A&M University Press.

Taber, S. W., and S. B. Fleenor. 2003a. Range extension, habitat, and review of the rare robber fly *Orthogonis stygia* (Bromley). *Southwest. Ent.* 29:85–87.

———. 2003b. *Insects of the Texas Lost Pines.* College Station: Texas A&M University Press.

Taylor, R. 1991. *The feral hog in Texas.* Austin: Texas Parks and Wildlife Department/Federal Aid Report ser. no. 28: A contribution of Federal Aid (P-R) Project W-125-R.

Teskey, H. J. 1990. *The horse flies and deer flies of Canada and Alaska. The insects and arachnids of Canada, part 16.* Ottawa, Ontario, Canada: Biosystematics Research Centre.

Tharp, B. C. 1935. Unusual plants of the Ottine area. Pp. 4–5 in *First Scientific Field Meet: Palmetto State Park, Ottine, Texas*. ?Gonzales, Tex.: Palmetto State Park, Tex. Board.

Tietz, H. M. 1972. *An index to the described life histories, early stages, and hosts of the Macrolepidoptera of the continental United States and Canada*. 2 vols. Sarasota, Fla: Allyn.

Tiner, R. W. 1999. *Wetland indicators: A guide to wetland identification, delineation, classification, and mapping*. Boca Raton, Fla: Lewis Publishers.

Townsend, C. H. 1884. Notes on some Coleoptera taken in south Louisiana. *Psyche* 4:219–22.

Triplehorn, C. A. 1972. A review of the genus *Zopherus* of the world (Coleoptera: Tenebrionidae). *Smith. Contrib. Zool.* 108:1–24.

Tuskes, P. M., J. P. Tuttle, and M. M. Collins. 1996. *The wild silk moths of North America*. Ithaca, N.Y.: Cornell University Press.

Tveten, J. L., and G. A. Tveten. 1996. *Butterflies of Houston and southeast Texas*. Austin: University of Texas Press.

Valentine, B. D. 1947. Cicindelid collecting in Texas. *Col. Bull.* 1:61–62.

Van Duzee, E. P. 1904. Annotated list of the Pentatomidae recorded from America north of Mexico, with descriptions of some new species. *Trans. Am. Ent. Soc.* 30:1–80.

Van Name, W. G. 1936. The American land and fresh-water isopod Crustacea. *Bull. Am. Mus. Nat. Hist.* 71:1–535.

Vaurie, P. 1955. A revision of the genus *Trox* in North America (Coleoptera: Scarabaeidae). *Bull. Am. Mus. Nat. Hist.* 106:1–89.

Vitt, D. H. 1994. An overview of factors that influence the development of Canadian peatlands. *Mem. Ent. Soc. Can.* 169:7–20.

Walker, E. M. 1958. *The Odonata of Canada and Alaska. Volume II*. Toronto, Canada: University of Toronto Press.

Walker, E. M., and P. S. Corbet. 1975. *The Odonata of Canada and Alaska. Volume III*. Toronto, Canada: University of Toronto Press.

Weaver, J. E., and R. A. Sommers. 1969. Life history and habits of the short-tailed cricket, *Anurogryllus muticus*, in central Louisiana. *Ann. Ent. Soc. Am.* 62:337–42.

West, M. J., and R. D. Alexander. 1963. Sub-social behavior in a burrowing cricket *Anurogryllus muticus* (De Geer). *Ohio J. Sci.* 63:19–24.

Wheeler, W. M. 1904. A crustacean-eating ant (*Leptogenys elongata* Buckley). *Biol. Bull.* 6:251–59.

White, R. E. 1983. *A field guide to the beetles of North America*. Boston: Houghton Mifflin.

Whiting, M. F. 1991. A distributional study of *Sialis* (Megaloptera: Sialidae) in North America. *Ent. News* 102:50–56.

Wilcox, J. 1936. Asilidae, new and otherwise, from the southwest, with a key to the genus *Stichopogon*. *Pan-Pac. Ent.* 12:201–12.

———. 1971. The genera *Stenopogon* Loew and *Scleropogon* Loew in America north of Mexico (Diptera: Asilidae). *Occ. Papers Cal. Acad. Sci.* 89:1–134.

Willemse, L. 1985. A taxonomic revision of the New World species of *Sirthenea* (Heteroptera: Reduviidae: Peiratinae). *Zoologische Verhandelingen* 215:1–67.

Williams, A. H. 1995. Adult female *Mydas clavatus* (Diptera: Mydidae) feeding on flowers in Wisconsin. *Great Lakes Entomologist* 28:227–29.

Williams, W. D. 1970. A revision of North American epigean species of *Asellus* (Crustacea: Isopoda). *Smith. Contrib. Zool.* 49:1–80.

Wolcott, A. B. 1908. The North American species of *Chariessa* (Coleoptera). *Ent. News* 19:70–72.

Wong, H. R. 1954. Common sawflies feeding on white birch in the forested areas of Manitoba and Saskatchewan. *Can. Ent.* 86:154–58.

Wood, F. E. 1962. A synopsis of the genus *Dineutus* (Coleoptera: Gyrinidae) in the Western Hemisphere. Master's thesis, University of Missouri, Columbia.

Yanega, D. 1996. *Field guide to northeastern longhorned beetles (Coleoptera: Cerambycidae). Ill. Nat. Hist. Surv. Man. no. 6.* Champaign: Illinois Natural History Survey.

Young, A. M. 1980. Observations on the aggregation of adult cicadas (Homoptera: Cicadidae) in tropical forests. *Can. J. Zool.* 58:711–22.

Young, O. P. 1982. Perching behavior of *Canthon viridis* (Coleoptera: Scarabaeidae) in Maryland. *J. New York Ent. Soc.* 90:161–65.

Zaitlin, L. M., and J. R. Larsen. 1984. Morphology of the head of *Mydas clavatus* Drury (Diptera: Mydaidae). *Int. J. Insect Morph. & Embryol.* 13:105–36.

Zaragoza, S. C. 1986. El genero *Distremocephalus* Wittmer en Mexico (Coleoptera: Phengodidae). *An. Inst. Biol. Univ. Nal. Auton. Mex.* 56:189–202.

Zimmermann, M., and J. R. Spence. 1989. Prey use of the fishing spider *Dolomedes triton* (Pisauridae, Araneae): An important predator of the neuston community. *Oecologia* 80:187–94.

Zona, S. 1990. A monograph of *Sabal* (Arecaceae: Coryphoideae). *Aliso* 12:583–666.

Index

Scientific names and pages with relevant figures appear in *italics*.